Crosscurrents North

Alaskans on the Environment

Crosscurrents North

Alaskans on the Environment

Marybeth Holleman & Anne Coray, Editors

University of Alaska Press
Fairbanks, Alaska

Second printing, September 2008

University of Alaska Press
P.O. Box 756240
Fairbanks, AK 99775-6240

ISBN 10: 1-60223-022-6
ISBN 13: 978-1-60223-022-4

Library of Congress Cataloging-in-Publication Data

Crosscurrents north : Alaskans on the environment / Marybeth Holleman & Anne Coray, editors.
p. cm.
Includes bibliographical references and index.
ISBN 978-1-60223-022-4 (pbk. : alk. paper)
1. Natural history—Alaska. 2. Natural history literature—Alaska. 3. Alaska—Description and travel. 4. Alaska—Environmental conditions. I. Holleman, Marybeth. II. Coray, Anne.
QH105.A4C76 2009
508.798—dc22

2007042900

Book design by Paula Elmes

This publication was printed on paper that meets the minimum requirements for ANSI/NISO 239.48-1992 (permanence of paper).

This publication was funded by the Terris and Katrina Moore Endowment.

Contents

Preface

ONE TIME RICHARD NELSON took me fishing for our dinner somewhere north of Sitka. "How deep should I drop the hook?" I asked. "Oh, until you feel it hitting the backs of the fish," he said. Sure enough, that's what happened.

My first reaction was the jaw-dropping awe of every visitor to Alaska: the sheer fecund abundance of the place. And my second thought was: *every* place was once like this. The mouth of the Hudson, on whose upper reaches I live, was once as swarming with life; Chesapeake Bay; the Gulf of Mexico. The first Europeans to reach Newfoundland found cod the size of canoes, and in such abundance they could simply gaff them from the sides of their boats.

Oh, Alaska is different. Different because most of us got there later, because its harshness and remoteness offered it some protection. Because the history that happened centuries ago in the Lower Forty-eight, the history that turned it into such a biologically impoverished place, is only now happening in Alaska. *And different because there are plenty of witnesses this time*. When it happened everywhere else, not many people (Native Americans excepted) were conscious of it as anything but progress. But Alaska, while it has plenty of people happy to bulldoze and drill and clearcut, also has plenty of people who see something different: a last-chance wilderness, a last chance for people to figure out how to earn their living in pretty much the same place that the rest of creation earns its living. A last chance for some kind of sane relationship with the natural world.

Many of those voices are represented in this remarkable collection: witnesses to the wildness and the abundance and the meaning that comes with beauty, witnesses to the threats that endanger it all. They are some of America's most powerful writers, in part because they can plant their feet so firmly and deeply in their home soil. And that is why this book is so important for people living down below, outside Alaska, back in the more settled world. We've been told by "official Alaska"—its oil magnates, timber barons, and their politicians—that it's just outside meddling by know-nothing Washington-based environmentalists that keeps them from all their cherished money-making schemes. But this book demonstrates beyond any doubt the depth and power of local opposition. Not sentimental opposition—Alaska is blessedly free of most kinds of sentimentalism. But clear-eyed, gritty, not-much-money-in-the-bank, that-bear-could-in-fact-kill-you devotion to their magnificent landscape. For many of them that landscape—its beauty and meaning—has become the defining fact of their lives.

Down here in the Lower Forty-eight our task is mostly different: the preservation of small patches of wildness, and the slow, steady restoration of those much larger patches that we damaged in our carelessness. That restoration is the work of centuries, maybe millennia; it will be a long time before Chesapeake Bay looks like Bristol Bay. But Alaska is crucial to that work as well. It provides us with the baseline, a reminder of what the real world actually looks like. And it provides us, at least for the moment, with the reassurance that humans are not automatically fated to subdue everything around them.

It is infinitely sad that voices like these were not raised to protect and celebrate all the glories that once marked this continent. It is infinitely sweet and powerful to hear them raised now, almost at the last moment and in almost the last place. May they serve as a compass for us all to steer by, wherever our corner of the still-lovely world.

Bill McKibben

Foreword

IT IS WITH SOME DEGREE of trepidation that I attempt a foreword for a serious anthology of poems and essays with an environmental bent. Not only is it presumptuous to attempt to probe the hearts and minds of those who composed them, my own abuse of the muse hardly qualifies me as a proper critic. Nonetheless, I am prompted to plunge forth for I, like the authors, am an environmentalist. Isn't everyone?

Unfortunately, for some extremists who would exploit without inhibition, branding one an "environmentalist" ranks but a half step ahead of child molestation in their lexicon. Conversely, in *their* fervor, some environmentalists seem to ignore that there are three realms to the environmental kingdom: the physical, social, and economic. Unless all are brought into harmony, ones who hold the scepter will dance a discordant tune played by those who would subordinate the first to the latter.

This anthology, I hope, will lay some stepping-stones to help bridge the gap between "extreme" developers and "extreme" environmentalists. As a guide, trapper, government hunter, commercial fisherman, and office holder with a degree in biology, I often found myself astride that gap with a foot in each camp's fire.

First as a legislator and later during my eight years as governor, more than once I stood within that gap attempting to field missiles hurled from either camp. Often bloodied in the process, I was able to fend off mortal wounds by shielding actions behind three basic principles to which none could take rational exception: (1) Did the proposal comply with existing environmental law? (2) Was it something that benefited all Alaskans, or but a select few at the expense of the many? (3) Did the majority of Alaskans want it?

In holding each proposal's feet to the fire of these three criteria, some mega projects, such as a huge hydroelectric dam on the Yukon River and a petrochemical plant requiring massive subsidy, went down in flames; others, like the Trans Alaska Pipeline, were approved, but only after several further environmental and economic adjustments were made to the original proposal.

While the environmentalists who demanded those adjustments were at first derided, it was later recognized that had they not prevailed, the pipeline might well have been an engineering nightmare. However, even this modern marvel of construction could have better met those three criteria had environmentalists even more prevailed and the tankering of oil, which led to the *Exxon Valdez* disaster, been avoided. Ironically, it appears in retrospect that other of the environmentalists' proposals, had they been adopted, would have also vastly improved the pipeline's economic impact upon both the state and nation.

During our state's early years often there occurred what I called "Battles between the buck and the biota." Recognizing that in such conflicts the former will almost always win, I came to understand the need to mesh environmental concerns with both the social and economic. While most folk gave lip service to concern for the environment, too many seemed willing not "to do it right, no matter what the cost" but to do it right *only* if the cost to them was nothing. Accordingly, in order to make the case for the environment, it became necessary to prove that failure to accommodate its concerns would incur economic costs voters would not tolerate.

My ensuing emphasis on economics prompted many environmental idealists to castigate me. My response was, "Those who could care less about the dickey birds will sit up and take notice at some economic turkey pecking at their wallets."

In Alaska, as elsewhere, the pendulum between environmental and economic considerations swings from one side to the other, often depending upon conditions beyond the state's control. For example, when oil prices plunge, environmental concerns against imprudent exploitation plunge with them. Conversely, when those prices rise, caution and constraint again are fashionable. Those in public office trying to hold the middle ground are likely to be clouted by that swinging pendulum, first on one side and then the other.

When I first came to Alaska in 1946, few concerned themselves with the environment. There simply seemed so darned much of it, few felt the need for caution. Our population of about seventy thousand, most in Anchorage, was constrained by lack of roads and costly air service. ATVs and snowmachines were yet unknown. Hence not too much harm was done, despite a frontier mentality that deemed most noble efforts to *tame* the wilderness, not treat it with consideration.

For examples, wolves, wolverine, and coyotes were considered varmints to be exterminated. They, along with seals, Dolly Varden trout, and even bald eagles had a price upon their heads (or, in the case of trout, upon their tails). Aerial hunting and use of poison were uninhibited.

While most such abuses have long since been corrected, many people still fail to recognize that sins committed on behalf of the environment are almost always correctable, while those committed in the name of progress and development carry hidden costs and scars that resist eradication. Accordingly, I preferred to chance erring on the side of the environment. And while such actions as repurchase of imprudently issued oil leases in pristine Kachemak Bay were roundly damned by opponents at the time, the Exxon catastrophe created public awareness of potential hazards and served at least to illuminate, and thereby hopefully avoid, some of the perils that lie within that aforementioned gap: a gap that often is actually no wider than the space between our ears.

I hope this anthology will generate a bit more candlepower in providing that illumination. For too often befogged in the hyperbole hurled from each side of that gap is the vision of a future to which most of us aspire: a future where rational resource development is balanced with rational environmental safeguards. Only through such balance can we assure the prime values that drew and adhered many to "The Great Land" are retained for future generations.

Jay Hammond

Introduction

My islands, you are my islands.
The sky over them in the morning today is joyous.
Just so is the morning of today. If I shall live henceforth,
let them be just so in memory.
—LINES FROM AN UNANGAN (ALEUT) SONG, 1840

IN MID-NOVEMBER, a full moon rose over the snow-covered Chugach Mountains, sharp peaks etched by the moon's white light and washed pink by the setting sun. We stood, a small group braving temperatures in single digits, to see that silver globe rise. At the familiar sound of raven wings sawing the air, I looked up to see three black silhouettes, stragglers returning to their mountain roost after a day scrounging the city of Anchorage for scraps. In the valleys surrounding us, hundreds of moose gathered after the fall rut, soon to disperse for winter. And less than a half mile down the trail from where we stood, a brown bear slept on top of a cow moose he'd killed earlier that week.

As we turned to go, my gaze was held by ice crystals on dried pushki shimmering in moonlight. Twenty years I've lived here, and I'm still astonished by all this wildness a few short miles from my house. I turned to my teenage son, who knows no other home, and said, "Look where we get to live. We're so lucky."

Alaska: named from the Unangan *Oon'alashka*, "where the sea breaks its back." Like the high plains of Africa, the Galapagos Islands, and the Amazon rain forest, Alaska holds meaning for people around the world, whether or not they ever set foot within its boundaries. The decades-old battle over oil drilling in the Arctic National Wildlife Refuge and the worldwide outrage over the 1989 *Exxon Valdez* oil spill illustrate how what happens in Alaska increasingly affects all Americans, and all the world.

With the largest remaining undeveloped and uncultivated lands in the nation, Alaska is the country's single most abundant natural asset. Within its 378 million acres are Denali, the most massive mountain in the world and highest in North America; the Yukon–Kuskokwim Delta, one of the largest intact wetlands in the world; almost all of the continent's active volcanoes; over half of the nation's coastline; and more habitat diversity than the rest of the United States combined. Alaska contains the world's largest bears, the world's largest remaining runs of wild salmon, and many of the world's largest national parks, national forests, and wildlife refuges.

But—as the *Exxon Valdez* oil spill, the most damaging in history, made clear—Alaska has the draw of both wilderness and economic potential. Well known for extractive industries, Alaska contains the nation's largest remaining reserves of oil and gas and supports large-scale mining, timber, and commercial fishing. Now, with the shrinking Arctic ice pack and endangered Alaskan polar bears, it is also the canary in the mine warning us about the most massive environmental disaster of all: global warming.

From its dual role as the nation's last great wilderness and primary resource colony, deep rifts arise among those who live here. Alaska is home to a diverse indigenous population striving to maintain ancestral traditions while finding a way into the twenty-first century, a last place of economic opportunity for "boomers" looking to get rich quick, the end of the road for rugged loners in need of unfettered space, and a refuge for those seeking a life whose rhythms align more with the natural world than with the urban-industrial world of modern society.

When I came to Alaska, I moved from a state where wolves were being reintroduced to the wild to a state where wolves were being hunted and shot from airplanes. The contrast astonished me then and remains instructive now—a kind of parable of the stark difference between environmental issues and values in Alaska and those in the rest of the country.

"How is it that you are such an environmentalist, being born and raised in Alaska?" asks a friend of *Crosscurrents North* coeditor Anne Coray, in Coray's essay "Precarious Preserve." It's widely assumed, both to those who live in Alaska and those who don't, that most longtime Alaskans have little concern for environmental issues. Any call for protection of Alaska's wildlife or habitat, the well-worn story goes, comes only from "outsiders." Witness policies such as state-sponsored aerial wolf hunting, or mining and trapping in national parks, or clearcutting virgin rain forests—policies that would be abhorrent in other parts of the country and much of the world—and it's easy to assume that a newspaper columnist was correct in recently writing, "The lower 48 has entered the information age. Up north, a dig it-drill it-chop it mentality still holds water."

As if to underscore this apparent view, all three recent candidates for governor trumpeted their support for oil development in the Arctic National Wildlife Refuge. In fact, no politician at any level of government in Alaska would have a chance at getting elected if they did not support such mega-development projects. Development is always seen as the best choice, no matter the cost to the environment. Recent projects and proposals have included a seven-mile, $80 million road to a town of three hundred people; massive dams on the Yukon and Susitna rivers; and an open-pit copper mine in the Bristol Bay area's biologically rich and remote wilderness. It seems that Alaskans see the land and sea as a treasure trove of resources waiting to be converted into cash, rather than as gifts that require sustainable use and preservation. Either that, or they think they can have it both ways: unchecked development and undiminished wilderness.

As the essays and poems in *Crosscurrents North* attest, not all Alaskans agree. The driving inspiration for this anthology arose from a desire to amplify the voices of those who are generally drowned out. These voices arise from a state where "environmentalist" is a dirty word, where Don Young, the sole congressperson, has called federal officials who support preservation "enemies of the state." These words spring from a place where there's more wild left than anywhere else in the country, but far fewer people raising their voices to protect that wild. *Crosscurrents North* stands as a protest document, providing another way of thinking about and imagining and living in harsh and beautiful Alaska.

It also demonstrates that the conservation ethic is not only the realm of those who have recently arrived in Alaska or those who come North for a few years of adventure. All of the writers in this volume have lived in Alaska for at least a decade; many of them, like John Haines, lived here long before the 1970s pipeline boom. Some, including Hank Lentfer, Libby Roderick, and

Susan Derrera, were born and raised here, and many have Alaska Native roots that go back tens of thousands of years, including Buffy McKay, Karin Dahl, and Joan Kane. Their environmental consciousness emanates from an intimate and abiding experience with their homeland.

These stories and reflections span the state, from Southeast's rain forest to the Arctic's tundra, from the Interior to the Yukon–Kuskokwim Delta and the Aleutian Islands. In order to best convey evocative, personal stories that advocate for preservation of Alaska by Alaskans, only essays and poems are included. With the exception of classic poems by John Haines, all the essays and poems have been written within the last two decades. These works traverse the range of contemporary environmental issues facing Alaska—from the ethics of trapping to the effects of global warming. Whether in praise or in lament, they shed light on what is at stake in Alaska.

The literary landscape of Alaska is replete with stories by visitors; these views from the inside provide an overdue perspective. The authors in this volume share stories that come from staying put; they ask questions that arise from living intimately with a place; they consider how to respond to the responsibilities and conflicts of living in an environment where natural forces still reign. Tom Sexton's "April" displays a reverence gained from his intimacy with the cycles of long, dark winters and brief, ebullient summers in his cabin near Denali. In "The Experiment," Nancy Lord draws on decades of walking the spruce forests near her home to reveal an intimate, visceral example of the ever-widening effects of global warming.

In "Continuing a Conversation on Place, Poetry, Love," Amy Crawford asks, "If you / and I know something of love / and of this place then / where is our courage, our praise / in bringing it to word?" Writers in this anthology have found that courage, singing hymns to archetypal symbols of Alaska, such as Denali, brown bears, and rain forests, as well as to particulars often overlooked by the casual visitor—morels, sand lance, and wild berries.

Throughout, these writers recognize their good fortune in living with such wild beauty in all its forms. "This is the place where your gaze / took hold, found words and moods / for words, where others will come / for your small sound when you are gone," writes Mike Burwell in "Your Land." As Mary TallMountain—the first in her Koyukon village to be adopted by non-Natives and raised outside Alaska—said in an interview, "I was born with the most beautiful land in the world. It's really stern and harsh, yes, but something else, a terrifically spiritual land, to me."

Underlying this praise, however, is a growing sense of diminishment.

Several years ago, I visited St. Lawrence Island, where my husband was assisting residents of Savoonga in developing a small-scale commercial halibut fishery. One of the elders, when asked about his catch, said that he'd kept one halibut for his own use. "What do they call it?" he said, "Sub—sis—tence?"

For him and those who have for generations lived off that which land and sea provided, the very idea of "subsistence" as separate from other uses was as foreign as considering the walrus apart from the ice, or the salmon apart from the stream. Indigenous cultures in Alaska have seen dramatic changes in the past fifty years—of increased human population and development, restrictions and infringements on traditional lifestyles, and a barrage of technology such as snowmachines and high-powered guns—that have forever altered how they interact with the wildlife and environment. The Savoonga elder's question is a reminder of just how fast the change has been.

As Walter Meganack's "The Day the Water Died" shows, Alaska Natives are increasingly confronted with the myriad ways in which modern society can—and does—destroy that which they thought was indestructible, like the oceans, like the ice pack. "Never in the millennium of our tradition have we thought it possible for the water to die," he writes three months after the 1989 *Exxon Valdez* oil spill. "But it is true." And, as he describes, environmental degradation ultimately leads to social and cultural degradation: "If the water is dead, maybe we are dead—our heritage, our tradition, our ways of life and living and relating to nature and to each other."

Given the rapidity of change for all of Alaska since the pipeline boom of the 1970s, many works in this anthology convey an attendant grief and outrage over what has already been lost and what may yet be lost. Sometimes the losses are sudden and painfully obvious, as with oil spills and clearcuts; sometimes they are slow and incremental, as with global warming and encroaching urbanization, at first visible only to those who have lived long and attentively with the place.

Some environmental issues facing Alaska are familiar to those in the rest of the country, as Susan Pope illustrates when describing the loss of parkland adjacent to her Anchorage neighborhood. Many issues in Alaska, though, are quite different—because of the extent of remaining wilderness, the economic potential of extractive resources, and the unique geographic position. Some of these concerns have been apparent for decades, but some are so new that no one knows how to respond. Large-scale oil spills, expanded offshore seismic testing, and the innumerable effects of a warming planet are exposed, lamented, and challenged by these writers.

"The confidence that Alaskans once possessed, that the wilderness could not run out, has begun to evaporate," writes Marjorie Cole in her essay "In the Shelter of the Forest." Nick Jans, in "Wolf Wars," echoes that sentiment as he tackles the decades-old statewide battle over aerial predator control: "Wolves, even unseen, fill up a landscape with wildness, define it. You'd think, after the mess we made elsewhere, that people would know better, learn to value the last places where large-scale ecosystems without boundaries exist, complete with the predators that define and shape them."

These writers clearly recognize their own roles in the degradation of the places they love. Of the Arctic National Wildlife Refuge, Karen Jettmar writes, "I bring people here to bear witness—to share with them its wonder and to convince them to become a voice against oil development. Yet, by marketing the wilderness experience, I contribute to a degradation of the very experience I am trying to protect." In the past, as Joanna Wassillie writes, "traditional Yup'ik culture did not have any serious ways to destroy our ecosystem." Given the rapid proliferation of technologies, she says that "somewhere along the way, however, our cultures have failed to make the necessary adaptation to our modern relationship with the environment."

These conflicting emotions of praise, grief, and culpability often lead the writers to recognize and act on their own responsibilities toward their homeland. Walking away from the North Slope oil fields and into undeveloped tundra, Glendon Brunk speaks of his newfound conviction: "I had to see my part in the scarring of Alaska, not always an active part particularly, more one of compliance by silence and avoidance." Sherry Simpson reaches a similar conclusion in "The Middle Ground," where she writes of a valley in southeast Alaska slated to be flooded: "I had been afraid to care about this place too much, the way you might hesitate making friends with a terminally ill person."

In a very real way, the heightened environmental consciousness embedded in this collection is a circling back to an approach that existed long before Europeans set foot here. As Athabascan elder Howard Luke points out in "Respect Gaalee'ya," "Respect is how we share with each other. Respect is 'don't destroy that thing.' People want to leave their mark there. That's not respect. Respect is leaving it just like you've never been there."

Alaska Native cultures, as Luke describes, have cultural traditions that provide the possibility of more sustainable, harmonious ways of living with the land—teachings that come through passing along stories. And so this collection of essays and poems tells and retells stories about Alaska that matter, stories that, like Dan Henry's essay, "Slouching Toward Deer Rock," illuminate how Alaskans can learn from the past, whether the lessons come from Chilkat legends or the transgressions of cruise ship companies.

In the end, this anthology raises more questions than it answers. Should trophy hunting and trapping be banned? Is catch-and-release fishing ethical? How much faith should be put in science to better understand and "manage" wildlife? Can development and environment coexist, and if so, how? While her sons and husband go bear hunting, Kaylene Johnson, struggling with ambivalence, desires simply to watch a wild bear that is safe from hunters. In so doing, she finds, if not a clear answer, some measure of solace: "At the dark end of dusk, the crunch of my footsteps on the trail made real my own presence here—a place as tangible and mysterious as the spaces of family and motherhood and the conflicted human world in which I lived."

Perhaps there can never be full agreement on all these issues, but solutions may be found in at least considering them carefully and deliberately. For it's the questions asked, and the ways they are asked, that ultimately illuminate how to live. As Peggy Shumaker writes, "Is my flowing / through the world / a fit gift? Have I nourished / more roots than I undercut?"

In "Wondering Where the Whales Are," Eva Saulitis dives deep into questions about knowledge: that which she learns from scientific training, from Chugach elders, and from her own extensive, focused time with the orcas of southcentral Alaska. Saulitis highlights the ambivalence between these ways of knowing the world. Learning from the humpback whales near Sitka, Carolyn

Servid suggests in "The Possibility of Witness" that salvation may be in what humans don't know: "I have come to rely on the comings and goings of humpback whales to ground me outside myself. I use them as checks against my presumptions, as reminders of my vulnerability, as markers that help delineate my own human boundaries."

These writers find answers not in the hard facts of science alone, nor in the brief view of a visitor. Instead, just as artist Paul Cézanne spent day after day in the presence of a mountain so that its essential nature might be revealed in his painting, the writers in *Crosscurrents North* find meaning and an enduring regard that comes from the long gaze.

Marybeth Holleman

Poems and Essays

John Haines

If the Owl Calls Again

at dusk
from the island in the river,
and it's not too cold,

I'll wait for the moon
to rise,
then take wing and glide
to meet him.

We will not speak,
but hooded against the frost
soar above
the alder flats, searching
with tawny eyes.

And then we'll sit
in the shadowy spruce and
pick the bones
of careless mice,

while the long moon drifts
toward Asia
and the river mutters
in its icy bed.

And when morning climbs
the limbs
we'll part without a sound,

fulfilled, floating
homeward as
the cold world awakens.

Mike Burwell

Your Land

This is the place that is given to you
because you have found it yourself

because you stayed long enough
to move your gaze fully to the south

because you returned at times
to accept the ritual: to care

and be cared for by rock
and bush, creek and bird.

This is the place where your gaze
took hold, found words and moods

for words, where others will come
for your small sound when you are gone,

to listen for the door you left
swinging, just out of reach,

on immense hinges.

Carolyn Servid

The Possibility of Witness

By escaping the human scale,
we become more fully human.
—Chet Raymo

THAT WAS THE WINTER OF WHALES. From October through January, the water beyond our house was marked by steamy spouts, dark backs, and wide, notched flukes. When I sat up in bed first thing in the morning I could look out to see them moving in Eastern Channel. All day long they punctuated the water—tall exclamations of hot, moist air, the rhythmic appearance of dorsal fins like a series of ellipses in their movements, the long phrase of their surfacing ending with a show of flukes. I would stand at the window with the spotting scope, watching—a single whale here, three others over there, two more heading into the channel, another heading out to open water, several others against the far shore. On a bright, still day, those towering breaths would hold a shimmering light; other days a strong wind chased them across the water. Each breath was followed by a broad back that would surface, arc gently, and sink down. I would get pulled into the rhythm of a single whale—the blow, then the back, arcing and sinking, again and again until the impulse for the deep dive. Then I could see the push from the center back that would begin the tighter arc, the black curve rising higher into the air, dorsal fin cresting the movement and raising the thick wedge of tail and the broad, winged flukes that stayed momentarily, then slid below the surface, leaving a flat, swirling circle of water and the silence of a marked absence. Over and over and over again this sequence, all up and down the channel, spouts appearing in so many places at once it was hard to decide where to focus. The whales pulled my days into a sequence that mirrored their surfacing. I would rise to watch them, stay with them a while, sink back into my work, and then rise again. At night I listened to the rumble and bark of their sharp breaths echoing like shouts in the darkness, and imagined their backs glinting in the faint light of an inky sky.

They came to Sitka Sound that winter in greater numbers than any time in recent years, some two hundred humpback whales. These animals are not infrequent visitors to the coastal panhandle of southeast Alaska. It has long been a feeding ground for humpbacks. Many have been documented—through photographs detailing individual markings on their flukes—as annual or seasonal visitors. One of their primary foods is the herring that school up by the thousands and move in and out of pockets along this coast—pockets such as Sitka Sound, a body of water that covers some 225 square miles. The congregation of two hundred whales, however, was out of the ordinary. Biologists knew that the Sitka herring populations had been increasing in recent years, and that the fish had begun to winter in Eastern Channel, a prominent passage close to Sitka. But no one expected that the increasing numbers of herring would attract so many humpbacks. That season Eastern Channel and Sitka Sound became a giant lunch box for those hungry swimmers and a spectacle for whale-watchers.

From my position on the sidelines, this direct correlation—more herring, more whales—seemed reasonable enough, a simple matter of fact. Beneath the surface, though, I came upon a host of questions that could not be answered with certainty. Why more herring? What conditions helped the population increase? Why did the fish change their wintering ground? How did so many whales know they were here? Biologists didn't know for sure. They have collected lots of information about the life cycles of herring and the habits of whales, and yet, necessarily, their knowledge is riddled with speculation. We will always be outside observers of other animals; some aspects of what we know must always rest on our best guesses. We watch the rhythms and cycles of nature the same way we watch the whales. We witness what comes to the surface, particular points in a larger encompassing sequence—points that are humpback and herring and season and number; points that, in relationship, outline a part of the greater order that contains us all. From our vantage point at the intersection of whale and herring and human, we try to understand all we can, and yet ultimately we wonder—at what we do know and at what we do not. What remains constant through all our efforts to explain, what goes undiminished, is the marvel of the entire construction—and of such a winter of whales.

I've lived much of my life at the edges of continents. A warm beach in western India introduced me at a very early age to white sand and saltwater, to the pound and pull of surf, and to oceanic expanses of water. My childhood perspective was a close and immediate one. I was fascinated with the myriad shells at the tide line, with cuttlebone and kelp, jellyfish and clams. I loved the froth of the breaking waves and the patterns of foam over the wet sand, loved to swim out and be lifted by those swells of water that rolled endlessly toward our long white stretch of shore. The vast seascape extending beyond the breakers didn't hold my imagination except when an occasional ship moved along the horizon or when the giant's eye I imagined in the distant lighthouse blinked brightly into the night. I wouldn't have thought to watch for the spout of a whale.

But the first suggestion of these animals came sometime during those beginning years. One of the earliest dreams I recall from my young sleep is of a whale at the top of the stairs in our house in that coastal Indian village. It was a blue caricature, a full rounded form tapering at one end to a forked tail. Its arrival on our stairs was not part of the dream. It was simply there, larger by far than I, filling the space between the wall and the top railing. My interaction with that whale took on the absurd reality allowed us in dreams—I poked it with a screwdriver and it rolled down to the landing where the stairs turned to descend into our living room. My mind left the dream at that point, but the whale stayed somewhere within me the rest of the night and rose into memory the next morning—and has stayed with me since. I don't attach particular significance to the dream except that it is my first recollection of an awareness of such creatures. Looking back, I wonder what experience—what picture or story now faded from memory—prompted my imagination.

Whales inhabit my dreams much more understandably these many years later. My imagination still lets in a degree of the absurd—they surface in mud puddles or approach me in the underwater basement of a house—but their form and behavior are true. They are humpbacks, flat-headed and knobbly nosed, with long, winglike flippers and longer tapered bodies, white pleated bellies, dark finned backs, and graceful curving flukes. Their exhalations tower above my sleep, their backs and tails move rhythmically through the watery scape of the dream.

I live now on Baranof Island in the Alexander Archipelago, snugged up against the northwesterly edge of the North American continent. Baranof is one of the largest in this three-hundred-mile stretch of islands, and one whose western coastline fronts on the North Pacific Ocean. The island's steep, forested shoreline is fractured and cleft by a litany of bays and inlets; its beaches are rocky. Intermittently there are stretches of sand. The shells, jellyfish, clams, and kelp here are the cold-water varieties of the North Pacific. The waves rolling in off the ocean get broken up by the smaller islands scattered within the broad reach of Sitka Sound. In front of our house, Eastern Channel collects the aftermath of that larger turbulence—sometimes the crowded chop of a storm, other times the slower, smoother swells that ease the water toward shore. The temperatures don't entice me to swim. Instead I ride the swells in a boat. And my imagination has been snagged by sweeping expanses of water. I am always watching for whales.

At the first sighting of a spout or back, I acquiesce. It is difficult to walk away from whales. The lure of their appearance is like the seduction of a game of chance. The possibility of witness keeps me watching, waiting, watching at the window. The stronger pull is to go out among them—to sit in my rowing dory and feel that burst and rumble of breathing in my bones, listen to the water swash and gurgle against that thick, dark skin, watch the rolling arc of back begin and continue until its twenty feet have passed me by and that final length, the heavy dripping flukes, has stroked the air with that enormous sweep. I drift amidst them perhaps foolishly, but my desire for proximity, for immediate experience of the whales, is more than mere curiosity or daring. It is rooted in the deeper connections we animals have with each other, in a common sentience. I cannot directly partake of the whales' sensibility. I cannot look a whale in the eye and share that moment of acknowledgment. But I skirt the edges of their world and the threat of their size and strength because I want a keen sense of their being lodged firmly within my own.

Watching whales from my living room was not something I ever imagined would be commonplace in my life. Going out among them in a rowboat was even further from my mind. For many of my years, whales were so remote as to be beyond belief. They were the blue caricatures of my young dream. They existed first in the pages of picture books, then encyclopedias, then guides to mammals, but not in the world I knew. When I was in college, Herman Melville made me believe in their power but didn't bring them any closer to my experience. It was later, when concern for their welfare began to put their images on T-shirts and in mail-order catalogs, that whales got locked into my imagination. I fell prey to the romanticized notions that pulled at public emotion. Their peril lent them a significance that made me pay attention. They were mysterious creatures, ethereal yet paradoxically more real. But my images were fantasies. I imagined transcendent beings, not animals. The enigmas that surrounded them—the unknown facets of biology and behavior—I turned into sacred mysteries. Popular recordings of whale songs reverberated through my conceits to create beings that harmonized with a greater power in the universe. My illusion seemed in the realm of possibilities. The chance to witness whales did not.

Shortly after I moved to Sitka, I met Chuck Johnstone, a skipper who has spent years navigating the waters of southeast Alaska. He and his wife, Alice, are longtime keen observers of the mountainous rain forest and coastline that stretch out from this community. Their comfortable forty-five-foot cabin cruiser, the *Fairweather*, has hosted many a charter to explore the inlets and beaches of these myriad islands, look for birds, and watch whales. A single trip on the *Fairweather* is enough to demonstrate not only the Johnstones' familiarity with the plant and animal life of this coastal ecosystem, but their sense of etiquette toward the larger natural world they live in. At the helm of his boat, Chuck is constantly mindful of the habits of birds and marine mammals as well as the hazards of navigation. When in pursuit of whales to watch, he moves the boat carefully and deliberately—staying the distance, slowing the engine, reading the signals of behavior to avoid irritation or interference, holding off or moving cautiously forward, letting the whales move toward him—all the time watching, gleaning more hints about these enigmatic creatures, all the time delighting in the close proximity to these giant mammals.

Chuck took me and a group of friends on our first whale-watching trip one December, a few chilly hours with three pairs of humpbacks in Katlian Bay just north of Sitka. Here was the opportunity I had assumed was out of this world. I approached the trip with both excitement and disbelief, wondering what it would be like to cross over from the imagined to the real.

My fantasies hadn't left room for what filled that stretch of day—steamy, rancid breaths, heavy black skin, ungainly flippers, massive, arching flukes—the truth of these animals, the tangibility of their bodies, the enormity of their motions. The whales surfaced again and again, near the boat, off in the distance, crossing over into our world each time they broke the water. I remember my friends singing to them, supposing they could call the whales to us, as though they were as interested in our presence as we were in theirs. We were enchanted captives, our imaginations embellishing all we witnessed. As we headed home I held those hours in my mind as one tries to hold on to something extraordinary, pure. But what I wanted to sustain was not simply the enchantment. I wanted those moments of contact to reverberate—the unobscured sound of that breath-shot, the exacting line of a rising back, the absolute clarity in the shape of the flukes. For those brief moments, the center of things had changed. It did not reside in me. It did not even reside in the whales. It rested in that space between us, that relation of living beings to one another. Our worlds came together at the point of breath. Together we partook of the damp, chill air.

I went back to Katlian Bay with Chuck and a friend three years later. January. Snow squalls moving through. We disappeared into those billows of white and lost all perspective. When we came out the other side of a squall, there was a startling moment of clarity, as though our temporary blindness had given us new eyes for a familiar landscape. The moment passed, and we recognized the frosted shore, rocks and trees limned with snow, points of land that told us where we were and where we needed to go. From Sitka, a simple route following the predominant curve of the easterly shore leads you directly into Katlian. The bay is shaped like a whale, an orca swimming east, its tall dorsal fin cutting into the landmass of Baranof Island to create Cedar Cove. The bay's entrance is right where the orca body would narrow before fanning out into a tail. As we motored up toward the head of Katlian, we found the whales—not orcas, but humpbacks.

A single pair this time, two adults swimming and diving side-by-side. They surfaced in calm water, shooting their hot breaths into still, hushed air. We were alone with them. Chuck slowed the *Fairweather* and eased toward them. The low throb of the engine rippled along the shore. The snows quieted everything. My friend and I stood on deck, talking to each other in muted voices, listening and watching—one blow and then another, one back and then another, parallel motions a few seconds apart, repeated three or four times before the deep dive. Chuck moved the boat to a comfortable distance from them and turned off the engine. The quiet settled in deeper. We waited on the deck without a word. Our eyes scanned the water, beginning at the swirling footprint where the whales had gone down and moving on the surface where we imagined they might move underneath. The dark water reflected riffles of the day's gray light. A soft lapping licked the sides of the boat. Our waiting stretched the length of the bay and back again, long moments for the cold to penetrate. We pulled up collars and tucked gloved hands in our pockets against the damp weight of that north coastal winter hanging close around us. What clues we had from the whales' dive were not enough to predict with any accuracy the next rising. Our patience had to be calibrated to the rhythm of those larger breaths.

Ffwhooooossshh! We turned just in time to see a blowhole go under and a dorsal fin rise not too far off the port side of the boat. The misty exhalation towered fifteen feet into the air. That whale was coming directly toward us. As its back sank slowly down, my friend and I began to get giddy. Now we could anticipate its next rising, time the several seconds between surface breaths. Ffwhooooossshh! Closer. The blowhole clear this time, the glistening wet skin stretched taught across the back. The dorsal fin peaked the arcing motion and sank again with ease. Our pounding hearts beat out the seconds and silenced all our words. Ffwhooooossshh! The whale had come within fifty feet of the boat. Its surfacing pushed its enormous body closer toward us. Did it know we were there? Would it veer off in time? We had caught its breath and held it in our lungs. The dark water at the side of the boat began to swirl. We leaned gingerly over the boat's rail and looked down. There was the broad beam of the whale's back slipping underneath the hull. We watched it disappear, then turned to each other, our eyes wide in nervous wonder. We turned again to Chuck inside the *Fairweather's* cabin. He was beaming his broad smile.

Human memory is a remarkable gift, but it doesn't always stretch to keep the details we would like. I have no recollection of the companion whale surfacing during those moments, but it may have. I know that we watched both whales again for several risings before we left Katlian Bay that afternoon. I know at one point they were swimming close to a rocky bank that climbed steeply out of the water into thick spruce and hemlock forest when a Sitka black-tail deer eased its way into the water and swam—only yards from the whales—around a small point of land to a less precipitous part of the shore where it could make its way uphill. And I know that later that night, when the full winter darkness had settled in, the aurora borealis spread a brilliant red corona across the cold dome of sky. When the hours finally slowed me down, I know that I reluctantly slipped into sleep, not wanting to let go of all that magic.

Watching humpback whales from a boat or from shore is like putting together giant pieces of a jigsaw puzzle. A spout, a flipper, a portion of the back, a fluke, a blowhole. My images of the full animal are provided primarily by underwater photographs and film footage or anatomical drawings, not firsthand observation. It is difficult to get a true sense of their size from momentary glimpses of particular parts—an arching back that comes only partway out of the water, a flipper waving solitarily above the surface, a set of flukes that, yes, are broad and heavy, but still just the tail of a creature I know to be forty or fifty feet long. And distance across water further diminishes the sense of scale. I have watched humpbacks breach—fly all the way out of the water, their tapered heads pointing their flight a certain direction, their flippers extended out like scallop-edged wings, their white bellies exposed to the sky. But a whale's forty tons do not stay suspended in air very long. Their return to the ocean is a spectacular explosion of water, and the cataclysm obscures the momentary clarity of that whale body in the air. Was it actually forty feet long? How big around? How big *are* the flukes, the flippers, compared to the rest of its body? Just where are the eyes positioned in relation to the blowhole and mouth? How far down the spine is that familiar dorsal fin? The sudden instant of the breach leaves me wondering if, in fact, I saw what I saw. The disruption of water is still evident at the surface, but the whale is nowhere in sight. And then a spout appears, a back, a dorsal fin. I revert to the more familiar, separate pieces of the puzzle, the more predictable rhythm of their appearance, however fleeting, and content myself. There is a fundamental elusiveness involved in watching these animals.

Some friends have a snapshot of me in my dory out in Eastern Channel. A steamy column of whale breath rises directly behind me to fifteen feet above my head. Dorik watches me through the spotting scope in our living room. When I come in he tells me of his nervousness when the boat became completely obscured by an arching back. I struggle to make this evidence real. For me, the breath-shot had little to do with size. It was that recognizable plosive burst of air, the position marker, the signal of something about to begin. I was absorbed in anticipating what would follow, in judging the direction of the whale's movement and its distance from me. When the back began to rise, my eyes tracked the curve of the spine and noted the irregularities in the shape of the dorsal fin, but didn't measure the height of that back above the water. How could I not? Perhaps a mesmerized human mind can only concentrate on a limited number of particulars; certain details inevitably get lost in the moment. Perhaps I knew, in spite of what appeared to be the case, that I was a safe distance from the whale, and its size was no direct threat to me. Perhaps one needs more intimate contact for comprehension to shift from something understood intellectually to something palpably real.

On Thanksgiving Day, that winter of whales, the brilliant sky was unlike the usual November one in Sitka. A glassy sheet of water stretched across Eastern Channel. The chill air was crisp and still. The reach of the channel a half mile beyond our house was alive with whale spouts, some clustered together, others spread out singly here and there. The sun made each breath luminescent as it hovered above the water. Dark whale backs glinted as they broke the surface. Flukes were sharp silhouettes against the bright winter light.

Dorik and I couldn't resist. He had not been out close to the whales, and I hadn't been in some time. We bundled ourselves up in long underwear, sweaters, hats, gloves, and life jackets, collected oars, camera, binoculars, and headed out to our dory. The boat floated quietly at its mooring, its red paint brilliantly lit by the day's brightness. I settled myself at the rowing seat, my back to the bow. Dorik sat in the stern, facing me and making room for my forward stroke of the oars. With two people and gear aboard, the dory's fourteen-foot length was quickly filled. I pointed the bow toward the far shore and eased into the rhythm of rowing. Dorik's eyes kept me on course as they took in whale risings beyond us.

Before long we were out in the channel and could hear the whales blowing. Each breath-shot turned our heads—this way, that, over here, over there. They were all at a fair distance, some moving up the channel toward Silver Bay, others heading the opposite direction toward open water. How many whales? It was hard to keep count; they were surfacing in so many places. If we followed a course straight toward the far shore, our path would likely cross one of theirs at closer range. My rowing stroke was steady. The calm water let the dory glide easily, pushing a slight, gurgling wave at its bow. The rhythmic bump of the oars against the wooden thole pins was noise enough, I hoped, to reverberate through the water and tell the whales of our presence. Whenever we stopped, I tapped my boot on the bottom of the boat to continue the message—we . . . are . . . here; we are . . . here; we . . . are here. It was a minimal and token effort, I knew. Friends whose boat is powered by an outboard engine—ostensibly enough sound to warn off a whale—had run full force into a humpback that surfaced unannounced immediately in front of them. And the whales made plenty of their own noise. Recent underwater recordings in Eastern Channel had revealed a cacophony of vocalizations: burbles and groans and yawning snores, single tonal notes and repeated yips, bubbling rumbles that descended from high pitch to low. These were not the acclaimed humpback songs recorded in warmer waters, but they were enough to override the ringing of a bell buoy or the whine of a boat engine one could hear in the background, let alone the tapping of a rubber boot against a wooden hull. Still, I felt safer with my own noise than with silence. Tap, tap, tap, tap. Tap. Tap, tap. Or the oars: thump and slide, thump and slide, thump and slide.

Out in the middle of the channel now, our course looked like it would cross the paths of two different groups of whales. A pair was swimming together out toward the wide mouth of Sitka Sound. Another group of five were headed the other way, toward the pair, toward the narrower entrance to Silver Bay. Whether or not the groups were in line with each other was hard to tell from our still somewhat distant vantage point. I kept rowing, trying to gauge our speed against theirs, trying to second-guess what the whales would do. Both groups seemed to be keeping a steady course. I tried to do the same, knowing

I could stop before we got too close, but wanting to share with Dorik that proximity I had experienced before. I wanted him to know how loud and steamy those eruptions of breath could be, how rancid and fishy that air from a whale lung, how thick and heavy the black skin, how knobbly the snout and flippers, how remarkable the pigmentation patterns on the underside of the flukes. When recollections of these images left me wondering in the middle of the night, I wanted my own witness confirmed.

The whales were down. We would have to wait now. They were still heading toward each other when they sounded, and we were verging on their course. I slowed the stroke of the oars, afraid we might already be too close. The glide of the dory smoothed into an almost eerie quiet. I started tapping my foot. We both scanned the water to either side of us for some signal. There were whales off in the distance, but the water around us was still. We spoke few words, as though the waiting demanded silence. My tapping boot seemed an intrusion, but I persisted.

Then it came, a breath-shot so loud it startled us both. A whale not more than thirty feet off starboard. Seconds later another shot. Another whale right with the first. I looked at Dorik and grinned. The two backs heaved themselves up slowly, parallel with the boat. The smelly mist wafted over us while the dorsal fins sank easily back down. As they slipped below the surface, eruptions began on the other side of the boat—one blow, then another, then three. The group of five was just a little further off port. This precise convergence—our being surrounded by seven humpback whales—was not what I had intended. My grin turned into a shaky smile. We sat still. My hands were tight on the oars, Dorik's gripped the camera. Whale breaths shot skyward all around us. Hulking, dark backs—close on starboard, off the stern, just beyond the port-side oar—followed each other to the surface, pushing the water aside. As one went down, another appeared—and another and another. The quick sequence turned into a confusion of whale parts and movements: blowhole, spout, back, blowhole, dorsal fin, spout, head, spout, back, dorsal fin. Our fourteen feet of dory began to feel too small amidst all those heaving forty-foot bodies. But they moved around us as though they knew we were there, and I translated that into a reckless confidence that it was all right to stay with them. Then they began to sound, arcing their backs high and raising their flukes. One went down off the port side, another started down off the starboard. Between them my eye got pulled to the water directly beside the boat. It had flattened into a swirling, jelly-looking mass, as though a strong current moved underneath. Any smile that might have been on my face instantly disappeared.

"Dorik, they're right underneath the boat!" I blurted in panic. "What should I do?"

"You know what to do," he said, reassuring himself, but not me.

I fumbled with the oars, as though I had never rowed the dory before. Panic shifted everything into slow motion. My arms felt like lead as I scrambled to establish a stroke. The boat only gradually began to move. I lost track of what the whales were doing—how many had gone down and where the others were. I knew instinctively they were not in front of the bow, and pulled blindly, moving us forward. By the time the dory gained speed, the water was quiet and we were alone.

The minutes immediately after are lost. I don't remember how I knew it was safe to turn the boat around and head home. It was only when we got to the beach that I realized we had rowed all the way back in silence.

We joined friends for Thanksgiving dinner late that afternoon, and our episode turned into a story, its immediacy fading with each retelling. But several nights later, Dorik sat up out of a fitful sleep, awash in an image of what might have happened. Days of my own uneasiness settled me at the corner window, watching through the scope. The comfortable dimensions of our house brought the disproportion to life. Our dory would easily fit inside with ten feet to spare; a single whale's head and tail would extend ten feet beyond each end of the sidewall. Adding mass to that length made me feel ridiculously small. Adding motion and water left me swimming. Adding six more whales made me leave the window to walk outside and feel the ground solidly under my feet.

Midway through the days of this writing a dream fragment surfaced in the night. A humpback broke the water and continued into the air, not in a full-flung breach, but as though its usual arcing motion at the surface simply took it a few feet beyond. In those few seconds of flight the whale looked me directly in the eye, as though acknowledging my watchfulness. Then it disappeared into the darkness of my sleep. When I awoke, I fixed the dream in memory not as a token of the possible, but as a reminder of my humanness.

My witness of humpback whales provides for me a distant measure of a daily existence that was once defined and shaped by the habits and migrations of other animals. These northern latitudes are rich with a history of such a way of life, some of it current, some confined to stories or the memory of a few living elders. Traditional Tlingit Indian place-names of islands, bays, beaches, and inlets around Sitka Sound reflect a culture whose survival depended on knowledge of these patterns in the animal world. The movement of salmon, sea lion, deer, seal, and halibut told people where to go and what to do in what season. Humpback whales were undoubtedly an indicator then, too, of the size of herring schools and where they were. They were signals of the condition of things, of what people would need to do to survive. While I don't have to depend on such signals for my immediate survival, I have come to rely on the comings and goings of humpback whales to ground me outside myself. I use them as checks against my presumptions, as reminders of my vulnerability, as markers that help delineate my own human boundaries. It is in these ways that they help me survive.

Two winter cycles have come and gone since that marked year, and the humpbacks returned in the same numbers. What seemed like a phenomenon has shifted to become part of a pattern defining itself. This winter's whale season may hold more of the same. The daily sightings began weeks ago. Today I glanced up from my writing out into the windy rain of a gale. Through the gray light, bursts of white appeared on the water's surface. They pulled me downstairs and held me at the telescope. Six, eight, ten whales were visible in the channel. A different rhythm today—repeated spouts and backs were interrupted by the side of a head lunging out of the water, followed sometimes by a hulking midsection with flipper waving, and then one lobe of a fluke. The whales were feeding, their lunges bringing them to the surface sideways. They swam quickly, aggressively, changing course with each surfacing as though chasing an elusive target. All around them the air and water were fluttering with hundreds of gulls. Feed of some kind had come with this storm, and the whales seemed caught up in the frenzy.

I stood inside at the window, protected from the rain and wind, determined again to be a witness. Watching, waiting, watching, I puzzled new pieces together, building on the familiar and trying to imagine the details I could not see. Hovering just behind my eye, in the center of my mind, was that recent fragment of a dream—the eye of the whale, watching me.

Tom Sexton

April

Another almost snow-
less winter, the stunned
earth unable to shed
its skin, when
a hushed sound
wakes you from
your restless sleep,
the first warm wind
of the new year,
frost rising
from the ground
lifting its coffin
as it goes,
leaving its seed
in the iris, so
this is what it means
to be holy, so
this is what it means
to be saved.

Hank Lentfer

A Voice for Shared Lands

IN THE DARK OF WINTER Anya and I flip through a pile of tattered charts and maps. Our fingers wander the coastlines and waterways of southeast Alaska as we dream of a spring trip. We usually go for a month, sometimes paddling, sometimes hiking. We start in April when the first of the migrants are winging north. It's chilly then, freezing nights, stale snow covering the beaches. It would be warmer in May but we can't wait that long. After the quiet stretch of winter our legs itch for the miles of beach and our arms hunger for the weight of the paddle. Besides, few other two-footed types are out and about in April. It feels like we own the whole place.

A few years ago, we opted for a coastal hike from Haines back home (one hundred crow miles north, twice that on foot). A stupid idea, really. The entire coastline is a jumble of boulders and cliffs. In the roughest section it took two days to cover ten miles. A photo essay of that hike would reveal a few hundred slides of various rock types with the rounded toe of a red rubber boot protruding into the lower corner. When the going gets tough the tough can't do a damn thing but look at their feet.

Yet, often, the effort of the journey breathes a vivid clarity into the memory of the trip. With blisters healed and legs uncramped, those weeks of hopping boulders and camping alongside streams blend with the richest days of my life. Vital days. Alive days. Days abandoned to the swing of weather and texture of stone. Days lured by the prints of wolverine and moose. Days ending curled beneath a flimsy tarp with my best friend, listening to the world.

Anya and I are lifelong Alaskans, both born and raised here. We plan to die here, too. The grave sites are picked out. Our love of place loops through our love of each other, growing more complex and defined with each pass. As we age, Alaska gets smaller. The rise in tourism, the spread of clearcuts, and the creep of roads have colored our childhood memories. It's hard to believe in the unexplored and undiscovered anymore. Our spring trips fill with a nostalgia for the past and dreams of the future.

I have developed an odd habit over the years. At each night's camp I fantasize about staying there for the rest of my life. I envision where to build the cabin to catch the morning sun. I evaluate the safety of the anchorage and look around for fresh water. I choose a garden site and study the tracks of my four-footed neighbors. The best part of the dream is it cannot come true. In the morning I gladly take down our tarp, kick apart the fire's ashes, and move on.

Our spring trips take place across a landscape that belongs to us all: millions of acres of mountains and glaciers, uncountable watersheds, thousands of miles of wild beaches, deep folds of forested hills. My home in Gustavus is literally surrounded by 3.3 million acres of national park. To the south stretches the seventeen million acres of islands of the Tongass National Forest. Scattered here and there are state marine parks and critical habitat areas. As long as our bodies allow, we will take spring trips and see but a fraction of our home ground.

My enthusiasm for the notion of sharing the landscape with humans and other critters grew with each mile hiked down that rocky shoreline. Trekking across private land required skirting homesteads and lodges to avoid the No Trespassing signs and barking dogs. Instead we nestled our sleeping pads alongside the prints of wolves and listened to the clear gurgling of streams and the gentle wash of waves. Not owning the land makes it all mine. And yours. And the moose's.

We carried a single page ripped from a book of Alaska topographical maps. The contours of coastline and litany of place-names on that wrinkled page remind me of the hike, and feed my growing gratitude that these places exist in such a wild state. Sullivan Island Marine Park, Endicott River Wilderness, St. James Bay Marine Park, Couverden State Park, Tongass National Forest, Glacier Bay National Park. Each name holds a story of someone's passionate persistence to protect a place. Environmental groups, both local and national, help with the lobbying, but the heart of each effort is traceable to one, maybe two, key individuals who love a place. Folks who know the saunter of individual bears, who can describe how the setting sun arcs across a particular ridge, who know when the crows first nested out on the point. Folks who view place as more than property and the world as larger than human. Folks who can infuse their love of place into the hearts of others. Folks who will not take no for an answer.

The passion and hard work of hundreds of such people, along with the foresight of President Jimmy Carter, resulted in the largest piece of conservation legislation the country has known. The Alaska National Interest Lands Conservation Act (ANILCA) was signed into law on December 2, 1980. The statute protected over one hundred million acres of federal lands in Alaska, doubling the size of the country's national park and refuge system and tripling the amount of land designated as wilderness. The patchwork of national parks, national forests, and wildlife refuges across this state bolsters my often faltering pride in America. By individuals speaking up for what they love, Americans created the means to share this earth on a scale unsurpassed by any other country. Studying a map of public lands is almost enough to make me go out and buy a flag.

Each year I celebrate the signing of ANILCA with a long walk in the woods. Alaska's congressional delegation, along with our former governor and many state legislators, curse the bill as a hindrance to commercial development. In today's corporate world, where privatization and profit are the golden rules, there is little tolerance for the notion of sharing. For two decades Alaska's politicians have drafted laws aimed at wiggling under, around, and through ANILCA. The growth of clearcuts, sprawl of oil rigs, and toxic seepage from mines are testimony to their success. Still, thanks to the dedication of countless advocates for public lands and Democratic control in Congress, most anti–public land laws never saw the light of day. This all changed when the Supreme Court inverted the political landscape in the winter of 2000.

I split two cords of firewood the night George W. was placed in office. I sat alone in my cabin as the experts discussed the court decision on the evening news. I clicked off the radio and sat in silence. Plagued by restlessness, I strapped on a headlamp and slipped on a wool coat. The creek gurgled through the dark as I made my way across the meadow to the hill of unsplit wood piled in the driveway. I centered a round on the chopping block in the feeble beam of light. Round after round cleaved beneath the maul. I swung through the winter night with a steadiness that made my heart pound and sweat tickle my ribs. I stripped to a T-shirt and kept swinging. After a few hours the batteries started to fade. I kept swinging until my light went out completely then stood, letting the cold night close in.

My fears that night have been surpassed. I knew it would be bad, but I did not imagine this much damage this quickly. A bill, crafted by our senior Senator Ted Stevens, now legally limits public involvement in decisions on logging in the Tongass. The Healthy Forests Act lays the groundwork for logging roads where they don't belong. Land swaps, trading acres of cut-over corporate lands for old-growth public parcels, are under way. The effort to open the Arctic Refuge to the ways of oilmen is relentless. Led by a man we did not elect, we, the public, are being banished from lands we supposedly own. CEOs sharing martinis and campaign contributions with politicians claim ever larger control over our parks and forests. The hard work of the folks who made my hike down from Haines possible is being undone under the banner of prosperity and profit. I am losing what I love.

The response to such loss? I could stock up on cheap whiskey and start drinking. I could shave my beard, buy a suit, and leave my cabin home to run for political office. Or I could simply roll up my sleeves and join my neighbors patiently working to protect three thousand acres of wetlands surrounding our little town. Since I don't care for hangovers or politics, the choice is easy. Bush's placement in office renewed my enthusiasm for our local land trust. The Gustavus Land Legacy, we call it. For over ten years we have struggled to raise the needed $3.5 million: tedious, slow work involving countless meetings with townsfolk, state politicians, and potential funders; dozens of days at the computer drafting purchase agreements and management plans, and writing grant after grant. The project is more complicated and time-consuming than I ever imagined. When I get discouraged I remind myself the toughest journeys leave the sweetest memories. And we're getting there. Over $2 million already in the bank, and more on the way. There is talk of what band will play at the celebration party.

Working with my neighbors has a twofold benefit. Thanks to our efforts, I will, as an old man, still hear wolves howl whenever I step out in the middle of the night to pee. The sandhill cranes will still stop and feed in the wetlands. The highbush cranberry patches will still grow their salmon-egg red jewels. Along with protecting the landscape I am preserving my sanity. The Land Legacy project gives me the daily opportunity to work for what I love. I know of no better antidote to despair. When I feel helpless in the onslaught of political news, I can turn off the radio, fire up my computer, and get to work. While public lands across the country are turned over to industry, my little arc of the world is going the other way. It's a tiny snapshot within the big picture. But I'll take satisfaction wherever I can.

We have a child now. Linnea will be six months old when we head out next spring. No hiking or paddling this year. We will choose a sheltered beach and stay put, some piece of public land that belongs to all of us and none of us. For Linnea's sake we will wait for the warmth of May. I dream of the three of us gathered beneath the tarp, Linnea learning about the eye-stinging qualities of a campfire. I see her shoving shells in her mouth and turning skyward at the sound of air through a raven's wing. I see her, each spring, roaming farther from camp to explore her surroundings. I see her return, giddy with the inexhaustible mystery swirling through the wild world.

Such dreaming feeds my gratitude for the work of people who loved this land before Linnea. People who toiled to keep the chainsaws at bay for a child they will never meet. Their successes and the wrinkled map of our hike down from Haines remind me that individuals can make a difference. They are the only thing that ever has.

I am an unabashed fan of public lands, be they seventeen-million-acre national forest or quarter-acre city parks. There is no greater gift for our children than a playground pungent and crawling with life. No greater gift to ourselves than a refuge big enough to make us feel blissfully small.

Linnea lies beside me as I write. She waves her arms, kicks her legs, gurgles, and grunts. A pleasant song about what she has come to know in her short life. A reminder to give voice to what I love. All we can do is sing.

Susan Alexander Derrera

August

This evening
as I rowed away
from the house,
my feet cool
under the collected
rainwater
on the bottom
of the boat,
I looked at the lake,
at how the rain
thrown across it
like children's jacks,
flashed
then disappeared—
and at that moment,
while rain curled
silver fingers
through my hair
releasing the wildness
there, I knew exactly
who I was
and what I loved,

and I thought of
you, whoever you are,
however lost you may be,
and I brought you
here
to listen
to the music

of the rain
on leaves and
the feathered backs
of grebes
and your own warm
skin.

When we rounded the island
the sky began to
lift and even the depths
were made clear—the smooth
gray rocks at the bottom,

your own
jeweled heart,

and after we were done,
tied off at the dock,
I brought you in
all wet and new
and offered coffee
in a small blue cup
and a piece of
rhubarb pie
hot from the oven,
the juices flushed
and running
on the plate.

Howard Luke

Respect Gaalee'ya

THERE'S A PLACE down on the Yukon River, below Tanana and Ruby, they call Bone Yard. You can see where those tusks are sticking out. Years ago, this place used to be a desert. We used to have sheep. This used to be just mountains here, not hills. All of the elephants and all of the snakes—some people say giraffes and everything—they had a big cave. The animals knew when the world was coming into the end there. They knew the world was coming in together. The world was going to change. That's what is going to happen right here, because we don't take care of our land. And that's what happened at that time. All of the animals got together, and they buried themselves. The animals told people not to go up there, not to go into that cave.

They had that big cave, and today, some places, you can find those tusks out there. When I was working on a boat, we used to go right by that hill and see those tusks sticking out. By the side of that hill, you can see them sticking about three inches out—just white. All of the animals said it was the end of the world, so they just buried themselves. This is why I say it's going to come yet. It's going to come because we're mistreating everything. We don't take care of it. People don't want to listen. They want to make that shortcut. That's why I say we've got to respect our land.

The animals told people not to go to the Bone Yard, otherwise something's going to happen. There are a lot of different places that people tell you not to go. People are not listening. They want to see for themselves. Someone told me a story just the other day. There was a fellow down by Ruby selling that ivory. He was picking it up, and he made a lot of money with that ivory, getting those tusks down there. He keeled over. So you see, there's something to it. When you're told not to do those things and you still do it, it's bound to catch up with you. This is why we're supposed to respect our animals. The animals used to be human at one time.

We've got these airboats, and the airboats are going out on the rivers and the flats in springtime, and they run over the eggs. Ducks have their eggs out there, and the airboats go anywhere—over the grass. They run over the eggs. That's why our animals are not coming back again. They've been mistreated. Those are the things I talk about, because they're really ruining our country.

Years ago, where there were birds' nests, they told us, "You can go look at the young ones but when you go to the young ones, you close your mouth. You don't breathe on the young ones, or else their mother won't come back." This is what we were told, and those are the things I want to put out.

We need to protect our animals, and protect our fish, and our ducks. I'm raising heck about the airboats. I've been doing that for the last year now. I wrote letters to the editor twice, and one lady gave an awful write-up about me. She said that the airboats don't make as much noise as the outboard motor. I sit in my cabin on the riverbank, and I can hear them twenty miles out there. What do they have those earplugs on for? Let them take those earplugs off, then they'll know what the hell the noise is. The airboats go out in springtime and they can run all over the grass, and they're running over the eggs. Ducks have eggs, and the airboats don't know where they're going. They just run anywhere it's wet. So they just kill all the young ones off. This one lady wrote a letter back to me. It's about time I wrote another letter and backtalk her again. Our animals are disappearing.

Every year we're getting less and less. Every year we're getting less water. We don't get that much water now. They predicted that years ago, that if we don't respect these things, they're not going to come to us. I mean they're going to disappear. As I was growing up, I heard people talking about it, and I'm still talking about these things yet. I believe what my elders told me. I'm an elder now, but I believe in what they told me at one time. And it's happening right today. We don't respect our animals and our trees.

Everything is alive. The trees are alive, the moss, the sand, the gravel. Everything is alive. The birch, we use all that. Years ago when my mother and them used to take birch bark off a tree to make baskets, they used to talk to the tree and leave something there. They said, "Always take some and leave some," and we are not doing it.

Just like the animals. The animals, it's the same thing. We have all kinds of animals. We used to have a lot of them. I remember way back in the 1930s, animals just started showing up. Before that we had to go hundreds of miles for moose. If we saw one track we just kept following it and following it. And we shared with one another. I don't care who it was, they got a little chunk. That's respect. And if that person says, "Thank you," that means he is giving you his wisdom. But now it's not that way. My own people are not doing it. I get moose, and I get fish, and I put it in the freezer. Beaver meat too, I got a freezer full of it, and I cook it up, and I give it to the people at Denali Center. It helps you when you do this, respect the animal. If you respect the animals, they will give themselves to you. I believe strongly in that, and I believe strongly about our trees. They're cutting our trees, and they're not using it all. They're just wasting the stuff. They're not taking every bit of it. They're not scaling the logs right. They're finally finding out that we're losing money on the trees. If they got someone that knows how to scale logs, it would be different, but now they're just clearcutting everything. It's not right, because everything lives on each other. There's our whole problem right today, that we're not living up to what we were told.

There's a story way back there: There was a caribou, and there was a bear, a grizzly bear, and he was hungry. He was hungry, so he took after this caribou. The caribou got away from him. The caribou looked back, and there was no bear. The bear was gone. So the caribou stopped and thought. He thought, We're supposed to share with one another. That bear needs. He's hungry. I'm going to give myself to him. So he turned around and went back. He went to that bear and gave himself.

Animals share and help each other by giving themselves to each other. They give themselves to us, too. Weasel is the leader. He's so fast, and he tells the animals, "Don't go there." That's the reason they call him Gaalee'ya, that's "luck." If we take care of things and respect them, the weasel gives us luck, and he tells the animals to give themselves to us. I think about those things. Right now it is just so disgraceful. We're just ruining the country. We are doing it ourselves.

They blame the wolf for killing all the moose off. They live on each other. They don't kill each other just for nothing. What are wolves going to live on if they don't kill moose?

I've said the same thing time and time again. They blame the wolf for killing the moose. I think we got more two-legged wolves running around here in this country than we got four-legged wolves. A wolf will just get so much, that's all. If you had your belly full, you wouldn't go back for seconds, would you? I wouldn't. I'd get sick. So that's what I say: If they want to keep the law real straight down like they used to do, they never blame the wolf. The wolves always take what's given to them, and then that's it. 'Cause there were wolves ever since I was a kid, and they share with each other. Animals all share with each other. And the coyotes all share with one another.

Right now they're hiring a helicopter to go out and shoot wolves. You know how much it costs to shoot a wolf? How much does it cost an hour for a helicopter, a couple thousand dollars an hour? Besides the cost, it's cruel to kill a wolf like that. The wolf that is hunted from the air is just like a man. He has no place to go, and that's just cruel.

Right now, moose season is open. I just come back from town, I saw moose antlers on a pickup right ahead of me coming down Airport Road, and no meat or nothing, just antlers. And they're going down to send it out. That's awful. People got to look into these things. What the hell do we have leaders for? We don't have leaders. People say they're our leaders, but they're nothing. They're just for themselves. I'm sick and tired of that. I hope that someone will listen to me and people will hear what I say and think twice.

But people got to work together. I get disgusted. Some people are just starving for meat. My own friends kill a moose, and they don't bring everything out of there. They bone everything out. Look at the head. We cook that up and use it for the potlatch. We have a big potlatch, and we never waste anything.

We've got to really work together to save our land. People used to say, way back when I was growing up, to really treat your land right—and your animals. If you respect those things, they'll pay you back in a lot of ways, but it's terrible how it is right now. It makes me sick once in a while when I see it. A couple of years ago, my nephew and I went out and saw a bunch of crows. I told my nephew, "There's something over there." So we went over, and there was a big moose. They had just cut the head off and taken that. We turned that in to the Game Commission, and they never did anything about it. Every time we turned something in they asked, "Did you see it happen? Who did it?" That's a hell of a law we got. One of these days, though, they shall see. The city is getting too big for itself, and people are getting too big for themselves. They are just for themselves. I got relations here in town. They got a moose, and do you think they cut me off a steak? No. I'm not going to tell them I need a chunk of meat. That's them. They got to decide that themselves. That's how it is right today.

A lot of people don't understand what respect means. Why do they get an education if they don't learn about respect? Respect is how we share with each other. Respect is "don't destroy that thing." People want to leave their mark there. That's not respect. Respect is leaving it just like you've never been there. If you blaze a tree, blaze a big tree, you kill that tree right there. It's going to have a scar. It's the same thing as your finger or something like that.

Like the animals. People just take chain saws and table saws and cut them up. That's not respect. That's what these hands are for. When I cut a fish, my right hand never gets bloody. This left hand does the work. If you let go of the knife, it gets all slimy and bloody. That's how people cut their hands. I never let go of the knife. That's what they call respect. And that way people will see you, and they will say, "Oh, man, that fellow really works hard to respect the animals and respect his hand." You see, you don't get blood on your right-hand side.

There are a few things I want to leave behind. No one has talked about this lately. There were a lot of sacred places years ago, when I was a kid. They got a place down there at Denali; down there at the mountain used to be a sacred place. Just the big people—that's all—could go in there, like the chief—were the only ones that could go in there. They had to go through certain people, counselors or something like that. They had counselors in those days, like second chief and first chief. They had to talk to these people before they could go over to Denali. That was a sacred place. Nobody could go in there. This is why a lot of things are not working, because they tell you not to go in there, and no one listens. They just go in there. This is why the whole world is not working right for us. I know I do a lot of that too. I just say, "Oh the heck with it. I'm going to go over there." That is why we have got a problem right today.

Over there, on the side of Denali, is a little hill, and there's a tunnel. The crow was the one that found that place. And he was the one that named all these birds, like the mallard ducks and teal ducks, the honkers, the swallows, the robins, and all that. They're all different. And just like clay, up there on the wall, are all the colors of all the birds.

People used to talk about it. I remember when I was a kid, they said that they didn't want anyone to go into that tunnel. Because if you go in there, the things you want to do are not going to turn out right for you or something's going to happen to you in years to come.

I am particular in ways, superstitious in a lot of ways. I believe it too, because I learned it down the trail. When I fell in the water, and we were right in the middle of the river with no one to get us out, we got out on our own. The current just went right around us. So, you see, there was something. So I've been forgiven for what I did. And I believe a lot of kids will realize that. Something happens like that, and they realize that, if you keep telling them these things.

So I talk to the kids a lot of times, and I put it this way. I believe our stories come first, if we want to teach kids and start working with kids. We've got to start with stories, two, three stories, anyway—about how people survive and how they heal each other. If kids listen to that, they'll say, "Oh, yeah, that's the way I got to go." And they'll turn the other way—not all of them, but, like I say, "If I help one, I help a thousand." I tell the kids all the time, "I'm here to give you my wisdom, and I want you to pass these things on."

That's what these things are all about. I know, the last two, three years, it's getting worse. It's getting filthy. People don't respect. They don't share with one another. There are mothers and daughters that don't get along. They predicted years ago that would happen, and it is happening right now. This is why I say we have to listen.

Stories, stories come first when you talk to people. You're going to leave something, your wisdom, with them. 'Cause years ago the animals used to be human at one time. That's why they say to respect the animals.

Amy Crawford

Continuing a Conversation on Place, Poetry, Love

Remember, over bends in the Canning,
you proposed this northland as too
personal to speak of well, more
intimate than sex? I agreed, we have
no honest words for what we know
by feel. But as we dropped off
the stone plateau west of here,
I thought to invent a poetry—
and you said poetry is too personal
and intimate. Then I suspected
you were afraid of love.

I'm leaning into my backpack
south of Red Sheep Creek. You know
the place, though you've never rested
here. No birds singing, the horizontal sun
scribing shifting light across cloud bellies.
I've walked my blood awake, and now
pre-dawn's first white-crown
whistles into gusts that stir
drunken spruce to hum.

You, far from fanciful, would only
think to know the North in certain

dimensions—by trapline, by plane—but
you may admit that the North
is your anchor, that by touching it, lying
in it, listening to its curves, as in
the dark with a lover—rise your truest notions.
And yet, what is love if you cannot
whisper a name, cannot walk with it?

I'm pausing here, bones sunken
in moss, to tender the place
to poetry, to name it—
yes, to call
it love.
To speak
through what I know.

This morning feels thick
and dim, unwoken.
In the silent lag of the storm,
there are only ideas, if you know
to have them, of sparrows
rising and tending the quiet
white-grass prairies, of the
limestone-cobbled coulees
flooding in the next hard rain.

This chilled air tastes dry—
and when the clouds soon come down
they won't drip like they're sighing
through a sieve but will shiver
off snow as a moose shakes
her whitened coat after sleep.

You once said nobody agrees
on the meaning of love or wilderness
but everyone submits an opinion.

Be it above Tintern Abbey, at Walden
Pond, or by Tinker Creek, these
opinions live on, by the page. If you
and I know something of love
and of this place then
where is our courage, our praise
in bringing it to word?

Come down and leave
a trace. Let your bones walk
their bones with my bones, walk
your ideas into something new.

Daniel Henry

Slouching Toward Deer Rock

Everyone living in a community with other people is inevitably a rhetorician.
—Edward Corbett, Classical Rhetoric for the Modern Student

Ritual

Cradling British or Russian arms, iron-tipped spears, and twenty-inch daggers secured by scabbard and buckler, men of the L'koot stay their posts against their Chilkat cousins on the meadow isthmus of Deishu. Musket fire has all but ceased on this fourth morning of battle; the prickle of high alert is replaced by anticipation of normalcy. Headmen of each faction emerge from a small house owned by neither. They still wear the whalebone shoulder pieces and bear hide mantles of war, decked by prominent cedar hats carved into killer whale and sea monster. Draped over the right shoulder of each, though, is a Chilkat blanket, emblem of peace. "Enough bloodshed," booms Chief Danawak's voice against the walls of ancient spruce that define the meadow's edges. "Six men dead on each side. The debt is repaid."

Ranks break and fade into the woods as warriors rush off to large cedar canoes that will take them to Deer Rock. An hour's paddle north to the head of Lutak Inlet brings men of two watersheds to the intertidal estuary of the Chilkoot River and, with this high tide, upriver to the rock. Early conflicts were waged a week ago at the river mouth, then nine miles south at Deishu, on the northern base of the Chilkat Peninsula before it forks the Lynn Canal into Chilkat and Chilkoot Inlets, halfway between two river territories.

The eulachon run on the Chilkoot was strong this spring, while the Chilkat produced only moderate numbers of the smelt whose oil is prized throughout the region. L'koot villagers also secured exceptional stocks of sockeye salmon, dried and smoked last autumn in such quantities that winter sheds are still more than half full into spring. Chilkats, whose three villages are accustomed to plenty, ache from a lean season on their river, named "salmon storehouse" for its usual abundance, and are angered by their relatives' unwillingness to share.

Resentment ignited at a forty-day party, a traditional event in which clans unite to commemorate the life of one recently deceased, usually demonstrated by substantial gift-giving. Relatives were outraged when a L'koot man refused his Chilkat cousin's request for a pack load of eulachon. Harsh words were exchanged. The cousins grappled, then one stabbed the other in the face, a heinous offense among Tlingits. Two days later, a band of Chilkats ambushed L'koots fishing at the mouth of the Chilkoot. Following the dictates of tradition, internecine war broke out to achieve a balance of honor. And now, as is custom, peace is reached at Deer Rock.

Men dressed in fine regalia beach their canoes on separate sides of the rock; the two groups gather a few hundred feet apart on the sedge-and-iris shores in the riverine avenue hemmed by giant Sitka spruce. Some men hold thick white shocks of bald eagle feathers. After each assembly exchanges messages carried by young runners, drums begin to throb somber rhythms and, one by one, dancers step forward. Choral voices rise like the fluffy eagle down flung from seal gut sacks and lifted above the river by invisible currents. Dance movements grow jerky as the drums become more assertive. A low murmur issues from the passing of water in clear blue-green channels between bear-sized boulders along this lower stretch. Drums thunder, then boom to a sudden halt. Dancers freeze. River noise rushes in to fill the void.

Dancers gradually retreat to their comrades, leaving two men to face each other from either side of the rock. From their respective groups advance the headmen. Each costumes his dancer with a deer-hide cape and an antlered headpiece. Warriors are transformed to deer; the jerky, exaggerated motion of the initial dance is replaced with elaborate circles of studied movement. Drumbeat is muffled, voices low. Peace is in the making. As eagle down wafts among them like the fat, lazy snowflakes of spring, troops sit wrapped in Chilkat blankets, some resting their temples on bent knees, contemplative. Drums are silent, movements of the deer are furtive—first one stalks the other around the rock, then roles are reversed. No sound intrudes but the whispering river, always the river. Drums and chants resound when the deer eventually embrace on the grassy flank between rock and rushing water.

At the conclusion, Chilkats are invited to feast in L'koot clan houses. That night, they gather around a fire pit to listen and respond to former enemies who stand forth with salutatory speeches on the virtues of the opposition. Because the right hand is used to brandish weapons, tonight the men will show peaceful intentions by eating with their left.

Home Fires

Scenes like these have replayed in my imagination since settling in Haines, Alaska, in the early 1980s. Located at the end of Lutak Inlet ten miles north of town, Chilkoot River and Deer Rock exert powers over me that grow with the years. I listen to the voice of the river as I crouch on its banks, journal in hand, or stand in chest waders with a fishing rod poised for a strike, bottleglass water twisting around me.

On the wall next to this computer is a black-and-white photo of my then-five-year-old son, Charlie Skyhawk Henry, leaning over a low metal sign in front of a boulder mass as big as a wall tent. *Guwakaan Teiyee*. Deer Rock. It haunts me. To communities bound by the razor wire of partisan politics, I ask, where is our Deer Rock?

Perhaps it is in the cycle of battle and reconciliation among tribal factions that we might find answers to broader questions about human conflict over natural resources in Alaska and the world, as well as protocols for peace. If we accept, as part of the equation, conflict between opposing groups over natural resource questions, are we willing to stand behind laws that protect human communities as components of intact ecosystems? How should we behave toward those who disagree with us? What words should we use?

I am wed to these questions. Born first-generation-off-the-ranch to a teacher and a nurse who fled to the suburbs when I was six, I grew up straddling a rural-urban divide. Rural meant farm chores, hunting, fishing, camping, and evening storytelling; urban offered school, debate contests, concerts, plays, and rush hour. While attending graduate school at the University of Oregon in the late 1970s, I discovered my academic passion in the rhetoric of natural resource conflict. For the next thirty years I pursued a fascination with what Wallace Stegner called the "unbroken double song of love and lamentation," sung by besieged conservationists against the destruction of habitat for which we are ultimately culpable. In that time I have come to see rhetorical engagement as less a battle than a series of evolving dance steps.

In a state known for obstreperous politics, Haines is legendary. It is a place of exceptional beauty and abundant natural resources for which "Hainiacs" take great pride. We are a town of twenty-five hundred righteous know-it-alls who have discovered something like paradise in this place. Meetings about human incursions into wild places—timber sales, heli-skiers, gold mines, cruise ships, parks, fish, bears, eagles, noise, water, zoning, roads—draw lively, contentious crowds. Let these words, then, stand as testament to the magnitude of my neighbors' disharmony over competing mythologies, the language we use to bolster our visions, and the manner in which our words reflect the rhetorical gunplay that sets Alaskan town halls ablaze. And let me suggest a possible outcome of conflict—peace.

If you rise straight into the air ten thousand feet from Main Street in Haines, you will more clearly understand the physical context in which this drama unfolds. Separated by a fractured range of peaks and knife ridges that rise up to five thousand feet from valley bottoms, the Chilkat and Chilkoot Rivers meet the sea at Lynn Canal, an oceanic channel cut into the glacier-cloaked Coast Range. Only a few gradual south-facing slopes and dendritic river valleys support forests for human and animal communities. As you slowly turn for the 360-degree scan it becomes apparent that in this vertical world of rock and ice, forests and meadows are slender ribbons of green. Habitat is a gift.

Through these waters surge salmon, eulachon, halibut, cod, herring, shark, seal, sea lion, and whales: humpback, fin, sperm, orca. Bird migration arteries along the coast into Lynn Canal and over the Chilkat Valley pulse with sandpiper, plover, robin, thrush, duck, eagle, finch, heron, goose, tern, owl, and swan, each following an imprinted route into the interiors of Yukon and Alaska, often the Arctic.

Up these routes, too, humans probed the territory. First came the Tlingit, some of whom settled in the twin river valleys teeming with fish. By Tlingit law, the people of each watershed "owned" and defended the resources of their respective territories. *At.oow* is the Tlingit concept of property that includes possession of traditional regalia, artwork, rivers, fish camps, forests, mountains, even stars. At the head of Lynn Canal, the two tribes controlled trade on the dual routes into the Interior. The Chilkats charged a toll for the passage of prospectors and traders trudging over the thirty-three-hundred-foot pass at the head of the Chilkat River. On the Chilkoot Pass, L'koots maintained a monopoly as guides and packers. Never forced to sign a treaty nor capitulate in battle, Tlingit leadership retains some principals of *at.oow* to this day.

In the dozen years following the United States' purchase of Alaska in 1867, a wave of humanity swelled up the coast, shot from Frederick Jackson Turner's pressure valve at the Western margins of the continent. In his 1892 essay announcing the closure of the country's frontier, the Wisconsin historian described the growth of a nation characterized in part by an edge where opportunists, pariahs, and dreamers could escape the withering reins of domestic society. Unable to reconcile the rise of regulation and refinement that marks civilization, these few spilled into zones of "perennial rebirth" where they devised ways to flourish with harsh landscapes and unorthodox neighbors. "Through these enclaves of edge-dwellers," Turner wrote, "America sustained her identity as Land of the Free." Like most rural Alaskans, the residents of my community have adopted that identity as a creed.

In November 1879, eco-apostle John Muir arrived at Yendestucke near the mouth of the Chilkat in a canoe also containing missionary S. Hall Young and four Stikine Tlingits. After listening to three days of preaching, the feared northern Tlingit requested a mission. In his journal, Muir declared his respect for the "venerable" Natives, but warned that "it is too often found that in attempting to Christianize savages they become very nearly nothing . . . they mope and doze and die on the outskirts of civilization like tamed eagles in barnyard corners, with blunt talons, blunt bills, and clipped wings." On an outgoing tide, Muir and his party departed southward, leaving behind a people forever changed.

While Native people struggled with transformation, Turner's pressure valve sprayed thousands of dreamers and derelicts past the Haines Mission to Skagway, gateway to the Klondike goldfields, at the terminus of Lynn Canal fifteen miles north. Though the parade of passing ships seemed at first to have little effect on those who watched from the shores at the mission, they signaled far-reaching change.

While most of the nation's attention in 1898 was cast toward the Klondike, gold was discovered in creeks thirty miles up the Chilkat from Haines Mission. By 1903, when the U.S. Army began building its first Alaskan post, Fort Seward, to protect the gold traffic, the rush had faded. The fort, built on the hill above Portage Cove, a half mile from the mission, remained active for nearly forty years. Chilkat and Chilkoot salmon runs attracted as many as six canneries, which served hundreds of fishermen, many of whom were local Tlingits who devised large funnel traps capable of cleaning out a salmon stream. The gold never paid out as advertised, and the fishery crashed.

With fish stocks depleted and gold rushing down to a crawl, Haines faltered. The town entered its Timber Age in 1939 with the arrival of nineteen-year-old John Schnabel from Klamath Falls, Oregon. Two decades later, fifty-year federal harvest contracts set off a cutting spree in the Tongass National Forest that virtually gave away old-growth spruce and hemlock to sustain mill towns like Sitka, Wrangell, Ketchikan, and Haines. Tugboats hauled massive log rafts up Lynn Canal to be transformed into chips for Korea and cants (square-sided logs) for Japan. What began for the Schnabels as a three-man operation grew to a payroll of 120. A second mill opened. Haines boomed, but a handful of residents began to question the impacts of massive clearcuts, especially when the state government pursued timber sales in the Chilkat Valley. "This alarmed a lot of people," said former mill owner Schnabel, "so we ended up with a long struggle."

Not unlike the clan wars of Tlingit tradition, loyalties flared, leading to symbolic or actual clash. After several Haines residents completed an ecology class in 1971, grade-school teacher Vivian Menaker and her husband, Ray, banded with other teachers and professionals to form Lynn Canal Conservation. "We thought somebody needed to say something," Vivian told me during an oral history interview in 1995. "The mills were getting away with murder." In a town whose recent identity was derived mostly from its status in the timber industry, such a group was sacrilegious. Meetings turned ugly. Worse yet, the harangue spilled onto Main Street. Businesses refused to serve conservationists. Along the sidewalks of our three-block downtown I witnessed midday shouting matches between sober adults gripping the hands of their children. In a local bar, I watched men smashing chairs on each other over timber harvest quotas.

As failing natural resource markets sucked money and jobs from Haines, a new tide of biologists, entrepreneurs, and telecommuters showed up, drawn by drop-jaw beauty and bargain-basement property values. Tensions grew.

John Schnabel's mandate to mow the state and national forests drew national attention in the early 1980s, when proposed state forest logging plans imperiled a stretch of Chilkat River habitat that sustains the largest gathering of bald eagles in the world. Colorado senator Gary Hart's proposal that it be designated as a national park elevated local animosities to a new roar. The closure of one mill and the shaky condition of the second further jacked up hostilities. When a middle-school teacher contacted the Department of Environmental Conservation in October 1989 about black smoke belching from the remaining mill, local resentment exploded. An angry group of mill workers rushed into the school during classes, intent on taking the teacher outside. I was in the hallway when the principal intervened as a Tlingit mill worker shouted, "If the mill dies, Haines dies," as he was being escorted out the door. In memory I still see the protest signs, reddened faces, indignant voices.

Ghosts of legendary battles still linger.

As resource liquidation slowed, interest rose in Haines's charm. In 1991, one year following the closure of the second mill, the community enjoyed its highest income in history with the arrival of Disney crews for the filming of *White Fang*. Cruise ship representatives and tour operators demanded their share of the scenery. Real estate values doubled, then doubled again. A few brave property owners called for zoning and planning while opponents branded them communists. Burgeoning tour companies attracted swarms of visitors into small planes, buses, rafts, jet boats, four-wheelers, horse-and-buggies, and helicopters, for a glimpse into a landscape primeval, unaware of their impacts on our rural community. We respond as a frontier people—protective of our freedoms, defending our Edge.

Great Land Battleground

Alaskans come to public forums as gladiators. We rail against those who would dispute an assumed right to live out the story of our design. Meeting halls in Fairbanks, Tok, and Hydaburg ring with the passion of locals in the throes of democratic theater. Combatants across a state 365 million acres large march into community arenas brandishing swords meant to hack opponents, then return to their enclaves for affirmation and encouragement.

Bugs, weather, isolation, breathtaking prices, long distance, and close quarters exacerbate peculiarities in community styles, uniting us as a cranky collective. Internal dialogue between personality and ideology is more pronounced when the agitated few among 680,000 total residents speak out in a public forum. Testimony often devolves into what rhetorician Kenneth Burke calls "agonism," or the tendency to believe that two opposing sides exist for every issue. State versus Feds. Rural versus Urban. Native versus Non-Native. Greenie versus Redneck. Godly versus Heathen. Sourdough versus Cheechako. Agonism fills the newspapers.

For most residents, Alaska is a combination of haven, hideout, or hell. With such a grand backdrop, it's natural to idealize the "Great Land" by wrapping it in dramatic regalia. We proclaim because we must. Burke recognized this human impulse to protect identity, or "Self," by performing on a dramatist's stage. It is when we convey to others an urgency to act that we discharge rhetorical blasts with a frontier audacity that Frederick Jackson Turner would concur is characteristic of edge-dwellers' "nervous, restless energy, dominant individualism, working for good and evil [with] that buoyancy and exuberance that comes with freedom." Clothed in the armor of our ideals, we are left with our love for a difficult place, and the difficult people who inhabit it.

Old-timers say that it wasn't always this way. Ideologies were set aside for the sake of survival. Back in the day, neighbors were more likely to share burdens—putting up fish, driving to town, burying the dead. Not so now. Technological innovations (cell phones, four-wheel drive, solar panels) enable new settlers to live with less reliance on the kindness of strangers, whom we once called neighbors. Land-use conflicts dictate the drama of the day, played out on community stages across the state. I've heard folks complain about the endless acrimony as they pack for emotionally warmer climes. Left behind are the stubborn few who'd rather lock horns than settle for Seattle. Through our endurance comes a sense of propriety, and the determination to fight for a mythic way of life on the Last Frontier.

Opinions tumble like glacial streams from the mountains of our beliefs; they rush down steep ruts to collide with other freshets on their way to acceptance, rejection, or evolution in the confluence of public perception. When antagonists scrap over Alaska's natural resources, we draw from our streams in the manner that Aristotle, Kenneth Burke, and Edward Corbett called rhetorical. In his seminal work, *The Rhetoric*, Aristotle underscored the importance of *ethos*, *logos*, and *pathos* to the outcomes of our efforts. Aristotle called *ethos* the most potent of the "modes of persuasion," as it reflects the character of the speaker, including trustworthiness, authority, and commitment. If a persuader lacks *ethos*, he might as well sit down.

In Alaska, length of residency determines much of a speaker's credibility. Unless an obvious cheechako (in which case you can argue from expertise), you are likely to preface arguments with the length of your relationship to the resource. Linking a resource-use argument with Alaska Native people can be effective. In the Arctic National Wildlife Refuge (ANWR) debate, Gwich'in people of Arctic Village have wielded their primacy as a formidable weapon against U.S. presidents and oil executives who would drill on caribou calving grounds. Conversely, Tlingit corporations have asserted longevity as a mandate to harvest old growth from the rich timberlands of southeast Alaska. Sealaska

CEO Bob Loesscher told me in 1994: "Tlingits never lost a war, never signed a treaty. We use our lands responsibly, and for the long haul. Non-Natives are here as our guests." An Anchorage-based Native corporation uses the slogan "Doing business in Alaska for 10,000 years."

For non-Natives, one's position on the longevity yardstick is determined by stark increments, especially when arguing about place. Of the fifty-four Haines residents that testified at a February 2005 hearing about the state's plans to build a road to a dead end sixty miles up Lynn Canal, forty-two included in each three-minute speech the number of years spent in the state. Of those who passed over the argument, I noticed that ten had lived in Haines for only a year or two. On top of one's badge of residency, Alaskans tend to be skeptical of Outsiders who have not walked the land in question. Our lone congressman, Don Young, has made a political career of castigating those who oppose oil drilling in the Alaska National Wildlife Refuge but have never visited the refuge. "Shame on you," he bellows. "How dare you." As I have often coached my debaters, people never make personal attacks unless all other strategies have failed. Bullying is an act of desperation.

Which argument is the most reasonable? What makes the most sense? To determine that, the advocate usually relies on logos, the assumed standard from which most legalistic or policy decisions spring. Numbers shock and satisfy, confirm and confound. Numbers are the main ingredient for the contention that sensible resource use should depend on demand, or the greatest good for the greatest number. Promoters have always hawked Alaska's huge potential as reason enough to develop it. Turn-of-the-century developer Charles Tuttle put his grand spin on Turner's thesis when he reacted to the 1910 United States' census denoting 100 million citizens: "The doors of Alaska, with its 586,400 square miles of new opportunities grandly are swinging open to the people of the United States and to the overcrowded countries of the world! Do we not see that old Nature never sleeps—never becomes weary?"

Today's congressional debates reflect a broader perception of Alaska as frontier to be preserved, not conquered. In the heated discourse leading up to the 1980 passage of the Alaska National Interest Lands Conservation Act (ANILCA), Ohio senator Jon Sieberling argued in favor of a parks bill that more than doubled the holdings of the national park system. He cited a Harris poll that "80 percent of the people were concerned with the deterioration of the quality of life in America, in all its aspects, and certainly the magnificent natural heritage we have received, and its destruction." Alaska senator Ted Stevens roared that the sample size of the poll was insignificant and ignored the majority of Alaskans who opposed the "land grab." Both sides flailed each other with the weight of Alaska's ponderous numbers until, in the final weeks of his term, Jimmy Carter signed the sweeping lands bill into law.

In the Great Natural Resource Debate, documents fill file cabinet fortresses, but for all the weight of evidence, declarations of the heart sometimes wield greater power. Pathos commands the attentions of audiences otherwise inured to charts, graphs, and pedigrees. Aristotle devoted a full third of *The Rhetoric* to detailed descriptions of anger, love, fear, benevolence, pity, indignation, shame, and other emotions that can affect the persuader's outcomes. Alaskan translation: When in doubt, shout it out.

"Anybody who says the ecology is fragile is an ignoramus of a goddamned liar. Anybody who tells me that this land was not put here to use is a socialist enemy of mine! Anybody who tells me trees shouldn't be cut, I'd use the axe on him!" When perennial gubernatorial candidate and ardent secessionist Joe Vogler made this promise in a campaign speech in the late 1970s, he crystallized the charged emotional rhetoric that flavors northern debate. By pledging themselves to long, dark winters, mosquitoes, and wilted produce, some Alaskans feel entitled to histrionics. For Vogler, whom I interviewed in August 1992, outrage led to founding, with Wally Hickel, the Alaska Independence Party that garnered, at times, substantial voter support. Vogler's fury soared as he watched the spread of the national "curse" of environmentalism: "If God actively curses people, I think they've had it. And it's just a matter of me livin' to see it play itself out." Ten months later, Vogler was murdered by mining partner Manfred West, over a business argument pertaining to Vogler's gold claim near Fairbanks.

In the emotional spectrum of Alaska land-use arguments, subsistence probably draws the most hot air. When it is a matter of food on your plate, you take it personally. Although the Alaska state constitution guarantees access to wild foods for all Alaskans, a provision of ANILCA requires an amendment stipulating a subsistence priority for rural Alaskans "in times of shortage." A quarter century, five special legislative sessions, and six governors later, lawmakers have resisted codifying the mandate that allows rural Natives the wild foods that Native lawyer Heather Kendall-Miller told me were the "essential elements" for cultural survival. Few arguments surpass genocide for raw emotional impact, especially when parents and grandparents speak for future generations. "We are a very law-abiding people," Yupik elder Myron Naneng said in a federal hearing. "But when obeying the law means that our children go hungry, something is wrong with the law." For this reason, Native leaders across the state argue for a subsistence priority, often linked with sovereign rule of tribal lands.

But special privileges based on ethnicity, warns equal-access lobbyist Dick Bishop, will not protect a dying culture. "The image of a person eking out a living by hunting, fishing, and berry picking is not true for many Natives," the retired state biologist told me. "To characterize it as essential to protect lifestyle is untrue." Subsistence advocates respond by comparing the average urban Alaskan's annual twenty-two pounds of subsistence food with southwestern Alaska villagers' six hundred pounds. In 2002, residents of Anchorage passed a referendum ordering lawmakers to amend the constitution allowing a rural preference in times of shortage. GOP lawmakers in Juneau refused to budge from the law, allowing equal subsistence rights to all. The battle wages on with enough experts and statistics to choke several polar bears, but pathos clinches it. In Alaska, linking ecological integrity with subsistence cements all three persuasive pieces. Arguing for future generations makes genetic sense, shows integrity, and can melt frozen hearts.

So. Why are we still raging over this?

It is the Alaskan Way, where the roar of voices, like glacial freshets, produces cacophony that can drown reason and wash away civility.

To the Rock

My hometown's scorched-earth reputation only changed when we discovered that we needed each other. As the new settlers built homes, families, and businesses, their stake in the community deepened. Some of the most cantankerous old-timers died still shaking their fists at anything that might hinder their freedom, like parks or zoning. A few reasonable leaders were elected. The economy took a dive. School attendance shrunk to half of what it was in Haines's timber heyday. The remaining residents took stock of themselves, and the reasons we could find to stay. For the most part, we laid down our heavy artillery. Whoever was left became more of an asset, too tough to chase away.

Our decades of struggle over local resource use taught us one important lesson: We are in this together. At his eightieth birthday party in 2000, former

Haines mill owner John Schnabel conceded the complexity of issues that once seemed clear-cut: "Over the years," Schnabel says, the rate of timber harvest proved unsustainable, "because the original thinkers never plugged into their minds the realization that not only does the forest provide timber for people, it provides habitat for wildlife; it provides a benefit to the fisheries; it provides aesthetics that so many of us like to see."

Eagles were our first teachers. Because the brawl over logging in eagle habitat resulted in national attention, a council representing all state and local interest groups was formed by then-Governor Jay Hammond to hammer out a solution before the federal government imposed its will. Included at the table were the sawmill owner, an environmentalist, a Tlingit leader, the mayor, a biologist, and others. Ultimately, it was an adamant refusal to accept Outside control that led feuding neighbors to forge the law creating the Alaska Chilkat Bald Eagle Preserve in 1982. In a speech delivered to a local audience fourteen years later at the annual Bald Eagle Festival, Hammond called the eagle preserve pact a "Mission Impossible" that turned out to be a "crown jewel in the annals of cooperative resource management." Three hundred people stood and thanked the governor with applause.

The cruise ship industry was our second teacher. After the sawmill shut down in 1990, local entrepreneurs cast their fortunes with the floating cities that passed Haines, bound for Skagway, fifteen miles north. Great white ships now daily disgorge as many as ten thousand summer passengers—purveyors of gold rush ambience—who stroll Skagway's boardwalks or ride the narrow-gauge train in search of frontier boutique. The Haines Chamber of Commerce yearned for a piece of the action. A dramatic decline in some salmon runs coincided with the arrival of Royal Caribbean Cruise Lines. By the following summer, investigation uncovered an elaborate scheme by the cruise company to dump human waste, dry-cleaning fluid, and photographic chemicals near the mouth of the Chilkoot River. The Environmental Protection Agency confirmed the connection and fined RCCL $7 million. A group of locals drew up plans and shepherded into law significant state and federal cruise ship regulation. "Fish kept us together," declared one business owner. "None of us could stop thinking about the fish."

At the horizon of the twenty-first century, a third teacher draws local attentions again to the grassy margins of a mile-long river. On the west bank of the Chilkoot River is a large rock near which summer tour buses squeeze for parking space, carrying thousands of tourists gawking for a glimpse of brown bears. A small sign identifies Deer Rock and describes its value for the original residents. Once a place for peacemaking, the rock bears the scars of ongoing battle.

A road building crew's dynamite in the early 1970s blew Deer Rock apart. Angry protest from local Tlingits convinced the Department of Transportation to reassemble the rock with concrete and return it to its original place. Chilkoot leader Austin Hammond, called Danawak to honor his forbears, led a ceremony to heal wounds and reassert the integrity of this place called the Chilkoot, where the river speaks an ancient language. The road brought its own lexicon, and new wounds.

Construction of a thirty-five-site state campground and a boat launch at the lake lured parades of recreationists. Several aboriginal graves were excavated and, after stern requests by Hammond, repaired. The river's roadside banks grew crowded with fishers eager to catch the muscular shadows that moved toward biological destiny. The state built a fish weir—a sort of iron fence across the river—so biologists could count and study returning salmon runs. Again, Hammond protested, but the weir remained. As human visitation to the Chilkoot increased over twenty years, so did bears. Summer nights, brown bears emerged from the thick forest to fish, sometimes to sample camp cuisine. Increasingly, human fishers competed with ursine fishers for prime river spots. Bears that became too comfortable with people were eventually relocated or, more likely, shot. By the late 1990s, strings of buses lined up along the riverside dirt road to disgorge thousands of cruise ship tourists seeking megafauna. More bears were shot.

Tensions soared among Natives, tour operators, recreationists, biologists, and bureaucrats, not to mention bears. Public safety and wildlife protection finally brought folks to the table. Problems were identified, personal commitments declared, baggage set aside. Through a web of common concerns, the Chilkoot Working Group was formed. Since then, plans developed by CWG have resulted in host monitors, parking enforcement, and a bear-viewing platform. Funny thing. Years of sitting together jawing about what's best for the community have turned fierce opponents into partners, sometimes friends.

Austin Hammond's nephew, Paul Wilson, once told me that Chilkoot people spoke loudly because they had to be heard over the river. And the river sometimes answered back with the voices of ancestors. "Always makes for quite the conversation," he chuckled. When I crouch for a journal entry in the cow parsnip and chocolate lilies on the banks of the Chilkoot, I swear that I hear them, too.

Amid fishing trash and cigarette butts near Deer Rock, I pluck eagle down from spruce branches, then walk to the water's edge and cast it into the air. Fluff wafts higher, then downstream, past the rock, in a river of voices. Steady, humorous, caring voices. No threats. No lambasting. Something like conversation.

Joan Kane

Dingmait

Black bird,
I have named you.

I have known
Your interdiction.
The otherwise

Sky was profuse
In its whiteness,
A confusion of

Snow come again
Into snow. Raven
Who would be made

Of wax, pitch.
In this way grow
Fast in my mind.

Shannon Gramse

Fall

Midway out on the McCarthy Road
we stopped to watch a pair of swans.
Their whiteness shone against the water,
the dying mountains, the alder and aspen.
I thought to tell you what I know of swans
at Niagara, how I've read about foggy
fall nights when whole flocks will descend
to the river to rest and regain their bearing
on Chesapeake and then the rapids stop them.
I wanted to explain how I've imagined them
thrashing and bobbing, fury of neck and wing
in thickening mist, and know why they fall
silent when the world turns white and roaring.
But then you passed back the binoculars and
said, "They're beautiful, aren't they?" So I only
answered yes and eased the truck into gear and
I held your hand as we rushed into the turning
like long-necked ghosts before an oath of snow.

Karen Jettmar

Finding Refuge

I LIE IN A TENT WITH TWO COMPANIONS, a hundred miles above the Arctic Circle, in the Arctic National Wildlife Refuge. Waking to the calls of redpoll warblers, I crawl out to greet the day. Across the river, I see him. Wolf. How long he has been there, I can't say. The Kongakut River sparkles in morning's fresh light; the black wolf stands by the water, watching me. I creep to the edge of the river on my belly. The wolf regards me. Our eyes meet. We stare, unmoving, the river between us.

I breathe deeply, tilt my head, and croon, my voice soft and low. The wolf raises his nose into the air and calls, voice rising to a high-pitched howl, then descending. I howl back. Pointing his nose down and ears back, he stares, and calls again, in a voice that is both moan and song. Back and forth, we call and respond; minutes become an hour. The long, low howls turn short and high-pitched, our voices trailing into silence.

The wolf's eyes turn from me, and he pivots away from the river. I watch him lope across the gray face of a mountain, his lithe, long-legged form disappearing among dark rocks. In the quiet, water flows over rocks, and blood pumps through my veins.

More than any other animal, the wolf is a symbol for wilderness. Mythic, mysterious, admired, and hated, wolves evoke strong emotions in humans, perhaps because they are so like us. Curious, expressive, loyal, and intelligent, they are wary of us, for good reason. Even in our largest, farthest north, and wildest national wildlife refuge, they are not safe.

The wolf's presence is a gift. Despite the Arctic Refuge's reputation as a northern Serengeti teeming with wildlife, an encounter with a wolf or any other animal is not a given, for the country is huge. While there is a seasonal explosion of life with the return of light in the late spring and summer months, and the simultaneous migrations of birds and wildlife, the land is spare, and wildlife sparsely distributed. In the case of wolves, when humans appear, they disappear.

My companions and I are the first people on the Kongakut River this year. In the coming days, there will be a dozen parties on the river, planes flying overhead most every day. There will be hundreds of ecotourists here to experience the Arctic Refuge they have read about in the news, on television, in tourism brochures, in appeals from conservation organizations. The dark wolf will retreat into a remote valley.

Come August, sport hunters will arrive to shoot animals. A single sport hunter can legally kill ten wolves every year in the Arctic Refuge, and five caribou every day. For individual Native hunters, the limits are even more generous: ten caribou per day and no limit on wolves.

Elsewhere in the state, the politically charged Alaska Board of Game wages an eradication program, both approving and funding an aerial wolf hunt. By killing wolves, the board argues, there are more moose available for people to hunt. Ignoring complicated coevolutionary process, and the integrity and stability of natural ecosystems, the board discusses expansion of the program throughout Alaska. "If you run 'em down in a helicopter and shoot 'em, that's pretty efficient," says Governor Murkowski. Even as Outside groups threaten a national tourism boycott, local letters to the editor proclaim, "Leave the tree huggers home. We don't need their money. Let's shoot some wolves."

Yet the threat of oil development and the specter of oil fields, pipelines, and access roads across the coastal plain of the Arctic Refuge loom as an even greater threat than the slaughter of wolves. Loss of wilderness is the greatest loss of all; once gone, it is gone forever.

In January 1987, I left my home in the small southeast Alaska village of Yakutat and traveled to Washington, D.C. Two other southeast Alaskans, Joe Sebastian and Joan Kautzer, from Point Baker on Prince of Wales Island, joined me at the Southeast Alaska Conservation Council's tiny office. We were there to lobby Congress on behalf of Tongass National Forest's ancient trees, swiftly falling under a massive taxpayer-subsidized fifty-year timber contract. For a week, we walked the halls of the Senate and House of Representatives, sharing with congressional aides our personal experiences living amidst the world's largest temperate rain forest. We tried to convince them of the folly of a taxpayer-financed below-cost timber program that built roads into pristine forests, while shipping the forest over to Asia, tree by tree, to be ground into pulp for making paper.

Coincidentally, Congress had begun hearings on a controversial environmental impact statement for the Arctic National Wildlife Refuge's coastal plain, or 1002 Area. Thanks to Alaska's pro-development senators, when the Alaska National Interest Lands Conservation Act passed in 1980, a provision was added, Section 1002, which mandated that 1.5 million acres of the coastal plain be evaluated for its oil and gas potential, as well as its wildlife and wilderness values.

After the mandated five years of field study, hearings were now being held around the country, including one in Washington. With Ronald Reagan as president, and James Watt as interior secretary, it was no surprise that the administration recommended oil leasing on the refuge's coastal plain. I'd never been there, but I'd been a backcountry park ranger in the Western Arctic national parks of Noatak, Kobuk Valley, and Cape Krusenstern. I decided to testify. It was my first time ever going before a congressional committee. I remember little of what I said, but my message was clear: Don't drill. Protect the wilderness.

The following summer, I traveled to the Arctic Refuge for the first time. Like most of bush Alaska, the largest and farthest north national wildlife refuge is far from any roads, so we flew there. Our party chartered a nine-seater commuter plane from Fairbanks to Fort Yukon, a Gwich'in Athabaskan settlement clustered near the confluence of the Porcupine and Yukon Rivers. It was mid-June; brilliant flowers painted the taiga, and mosquitoes buzzed in the air.

From there, we chartered a small bush plane to carry us still farther north. By air, we traversed countless lakes and ancient oxbows. Subtly, the terrain began to change. The Arctic Refuge lay below. Stands of spruce stretched up the valleys into hills engraved by caribou trails. We soared over a procession of rugged mountains and climbed over the northernmost extension of the Arctic Divide, that snow-covered backbone of the Brooks Range where the rivers change direction. Streams born of glaciers now sped north to the Arctic Ocean. Below us lay the Hulahula River.

Our plane settled in an alpine valley, landing on a gravel bar beside the river. After we unloaded rafts and food enough for a couple of weeks, our pilot took off in a cloud of dust, and we were on our own. I wandered off and climbed a small hill beside the river. Sharply pitched mountains rose around me. Glaciers nestled beneath their summits, and creeks ran into the river like quicksilver. A dozen Dall sheep climbed single file along a contorted, multi-layered limestone ridge. Two rock ptarmigans flew low over the tundra; in the silence between their low-pitched calls, I heard their wing beats.

For ten days, we explored the greatest true wild country left in the United States, a vast ecosystem of more than nineteen million acres, encompassing every arctic ecoregion—boreal forest, forest tundra, mountains, arctic tundra, and coastal marine—in its 250-mile breadth. We traversed deep-cleft canyons, spacious valleys, verdant foothills, and a vast prairie, nearly reaching the Beaufort Sea.

On the first day, we leave the river early, hiking along Kolotuk Creek. Our path lies on polished granite stones, intermixed with brilliant wildflowers. We cross the creek with little difficulty. It's covered with thick ice shelves (aufeis) split by meltwater. We step over deep, cold runnels, dodging waterfalls that pour into the creek.

Our daypacks bulge with fleece jackets, pants, sleeping bags, water bottles, and food as we climb up tundra hillsides covered with purple lupine and magenta Lapland rosebay. The land is saturated with water, as last winter's snowpack melts under twenty-four-hour daylight. We slog up wet tundra, pausing often to drink at the edges of the melting snow. We emerge to a flat tableland covered with shattered limestone. The tall peaks of the Romanzof Mountains lie just across the plateau.

It's difficult to grasp the scale of such a landscape. From far above the U-shaped Hulahula River valley, our tents are barely visible beside the river. The country feels untouched, as fresh as creation.

The sun stays with us, circling directly above at midday and spinning north across the horizon towards evening, as we cross Esetuk Glacier and continue climbing. At midnight, we stand atop a precipitous spur of Mt. Michelson, nearly eight thousand feet above sea level. Mt. Chamberlin, the tallest peak in the Brooks Range, rises to the west. The Hulahula dwindles into a tangle of channels glinting in the sunlight. We can see for fifty miles, maybe a hundred—and there are no roads, no settlements, no facilities. The ocean seems like a mirage, frozen as far as we can see.

Next day, hiking back to the river, I spot a blond form across a swale. It's a grizzly. Heedless of our passage, it plows into the meadow with its claws, throwing up clods of dirt. We continue down the ridge.

On the fourth day, hiking up a side valley, we wait for a red-backed vole to peer out from a cluster of dwarf birch. This time of the year, all that remains of its snow burrows are skeins of brown droppings and tiny bunches of dried grass. The vole darts from shrub to shrub, and pauses before disappearing into a jumble of rocks. My companion captures it on film, while I photograph lichens.

On June 22, the day after summer solstice, the tundra bares its secrets in the bones and remains of plants. I find the chalk-white skull of a caribou, and a bleached antler poking up from deep moss that has nearly buried it. A close examination reveals the tooth marks of a rodent.

Two days later, canyon walls rise steeply across the river, and whitewater rushes over huge boulders. Castle rocks and buttresses remind me of the Southwest. I want to linger, to explore every nook and cranny; we're traveling through so fast. The creek roars down an incised canyon, carrying the glacier melt.

We pack up and load the boats for the trickiest whitewater on the river. There are many rocks and big holes. The river, now deep in a canyon, makes a sharp right turn, followed closely by a sharp left turn. We follow a left course, negotiating around several rocks, pulling hard away from the sheer canyon wall. Our raft skims by a ledge drop. Water splashes over the gunwales, and just as we emerge from the last twist of the canyon, it begins to hail. Hailstones beat down upon us, blanketing the river. We scream as they strike our hands and heads. I surrender, hunkering down in the raft, only dipping the paddle now and again to keep us on course. The Hulahula writhes; hail swirls into the current.

In ten minutes, it's over. Sun bursts through a fissure in the clouds, and the hail melts into oblivion. We pull into an eddy, jump ashore, and flop onto our backs. Like mountain avens, whose parabolic petals absorb and reflect sunlight, our bodies soak up the sun's warmth.

Farther down the river we scrape over shallows, and days later, watch the river rise seventeen inches, churning mocha brown in the aftermath of an upstream squall. Fog moves inland from the coast, and we paddle through thick shelves of ice, listening to the crash of ice hitting the river and feeling surge waves beat up against the raft. We crawl on our bellies with hand lenses, marveling at the intricacy and delicacy of the tundra. We slip into deep pools of water, where fish dart among polished stones. Crouching animal-like in the tundra, we watch a small herd of musk oxen graze, a young calf protected in their fold. From high vantages, we gaze upon an elemental landscape, the creation of millions of years of patient evolution.

Our journey ends just inland of the Beaufort Sea, close enough to taste and feel its salty chill in the wind. For ten days, we never see another human.

For nearly two decades, I have returned to lead small groups of travelers in the refuge. Herein lies the conundrum. I want to protect this place. I bring people here to bear witness—to share with them its wonder, and to convince them to become a voice against oil development. Yet, by marketing the wilderness experience, I contribute to a degradation of the very experience I am trying to protect. Guiding services, advertising, photography books—each of these informs, yet somehow tames, the experience of the wild.

Consider the marketing blurbs of a few of the thirty-five recreational or sport hunting guide companies authorized to conduct commercial activities in the refuge:

"The refuge is among the world's last truly pristine wild places and one of the largest sanctuaries for arctic animals."

"Experience one of the most remote, magnificent wilderness areas on our planet."

"The land is so vast and untamed that salmon swim undammed rivers and caribou may live and die without ever seeing a human."

We draw people here; we create a demand for the "wilderness experience." A U.S. Fish and Wildlife manager related to me a conversation he'd had with the director of an international adventure travel company who spoke of the company's plan to "increase their product line." We live in an age of exploitation, where even wilderness rivers are a mere commodity.

More than fifty years ago, National Park Planner George Collins came here and concluded that being here made you "feel like one of the old-time explorers, knowing that each place you camp, each mountain climbed, each adventure with boats, is in untouched country." Now I worry that we are trampling it.

Last summer, I led a group of Great Old Broads for Wilderness down the Hulahula River. At journey's end, we arrived in Kaktovik, an Inupiat village at the northern edge of the Arctic Refuge. A red Twin Otter landed on the gravel runway; out jumped a magazine journalist and a photographer from *Condé Nast Traveler*. The pilot and his wife emerged as well, along with the photographer's wife and a guide. The journalist fired off questions at our party, while the pilot's wife railed. "We counted sixty-eight tents on the Kongakut River," she said. "I thought this place was a wilderness. It's getting so crowded down in Wrangell–St. Elias; tourism is ruining the park. We pioneered all those hikes that everyone's advertising now. We came up here thinking there'd be some space. Who are all those people?"

I commented that the Arctic Refuge has been in the forefront of environmental news for quite a while, particularly under the Bush administration. I asked the journalist about his story angle.

"Well, we're interested in the controversy over oil development," he said.

"But you write for a travel magazine."

"Sure, it's a travel piece," he said. "We intend to get the full experience. Anyway, our readers will never come here—too many mosquitoes."

The guide interjected. "Do you know any good places to land a plane to go camping where there's lots of wildlife and there aren't any people?"

It's a conundrum of the grandest proportions. The same energy that creates havoc with the environment is a source of our transportation to the very environment we are trying to protect. We sell a wilderness as though it were a theme park, complete with herds of caribou, grizzly bears, and wolves. Glossy travel magazines glamorize spendy Arctic Refuge camping trips, featuring fresh salads, steak, and wine. Every major outdoor magazine publishes articles about the Arctic Refuge. People hunger for something genuine, yet we destroy the very opportunity for confronting mysteries of self-discovery. There are too many of us here; the wilderness is getting crowded. According to a recent Harper's Index, the current human population exceeds the earth's sustainable carrying capacity by a factor of 1,000. The irony gnaws at me.

Just thirty-five years ago, Alaska's Brooks Range and North Slope were an unbroken continuum of *de facto* wilderness, without a road penetrating the fragile tundra. By 1975, the eight-hundred-mile Trans-Alaska Pipeline had been built. Traversing the state from Prudhoe Bay to Valdez, it crosses three mountain ranges and tunnels beneath 350 rivers and streams, a denuded swath of land that runs down the entire length of the largest state in the United States.

"There is a tension in the fabric of wilderness," wrote David Brower, "and when it is cut [it] withdraws fast and far." Like a slash in a tree that gets deeper and wider with each swing of the logger's tool, so have the inroads of technology cut deeper into the core of Alaska's wilderness.

People say the Arctic Refuge is remote. The notion of remoteness is an artificial construct in the twenty-first century. We have the technology to go wherever we want. A road runs the full length of the state, alongside the Trans-Alaska Pipeline. North of the Arctic Circle, it borders the western boundary of the Arctic Refuge. You can hike right off the road. If you have the money and the time to plan your own adventure, you can charter your own bush plane into the refuge. With more money, you can sign on to one of the many expeditions that adventure companies organize. Alaska's wilderness is accessible. Residents of the state have more bush planes per capita than anywhere in the world. "It is the expansion of transport without a corresponding growth of perception that threatens us with qualitative bankruptcy of the recreational process," wrote Aldo Leopold. We need to practice restraint.

Western civilization seems blind to the rapid environmental change occurring on the earth. People think Alaska is one big wilderness, intact and virtually roadless. Compared with just about anywhere else, it's true. But a lot of this open space is accessible; the land is crisscrossed by off-road vehicles, snowmachines, motorboats, planes, and helicopters. Tourists are amazed at Alaska's size and grandeur; they don't see the incremental loss of the wild. The newness, the spaciousness, overwhelms the senses. But change here has been rapid, the losses immediate.

Given a land so endowed with spectacular natural beauty, you'd think all Alaskans would be environmentalists, fighting to protect wild country. Instead, they are among the most pro-development citizens in the country. They see Alaska as a land of endless riches, theirs to do with as they please. Alaskans enthusiastically endorse resource development, ignorant of the consequences. The average Alaskan collects an oil revenue dividend from the Permanent Fund, thinks we should "drill ANWR," and travels to Hawaii for vacations.

Alaskans want every cookie in the jar. Environmentalists like me are labeled "extreme" and "enemies of the state." Political candidates shy away from association with the environmental community, for fear of sinking their chances of election.

Even many Alaska Natives support oil development, though this is changing. The Alaska Federation of Natives, Alaska's largest statewide Native organization, still officially supports Arctic Refuge development, though several Native groups officially protest this position. To the more than two dozen Gwich'in Athabaskan villages scattered across northeast Alaska and Yukon Territory, the Arctic Refuge calving grounds of the Porcupine caribou herd, is "the sacred place where life begins." Gwich'in caribou culture has existed for as long as people have lived in the northern hills, valleys, and lake country of the Porcupine and Chandalar Rivers—more than ten thousand years. The Alaska Inter-Tribal Council's membership of more than two hundred villages supports the Gwich'in position, opposing oil development in the refuge.

Though three decades of oil development on Alaska's North Slope have brought wealth and modern technology, increasing numbers of Inupiat are questioning the oil industry's subtle invasion and eradication of their traditional culture. For the Inupiat village of Kaktovik, located next to the Arctic Refuge, the oil infrastructure has always been 150 miles distant at Prudhoe Bay, while oil dividends brought a prosperous economy, with new homes, community sanitation facilities, vehicles, and boats. However, state and federal offshore oil leases press ever closer to Kaktovik. In the fall of 2006, oil giant Shell began seismic testing in the ocean off the coast of Kaktovik. During the first big autumn storm, some Bering wolf fish, normally found in the Bering Sea (and never seen in Kaktovik), were found washed up dead on the beach in Kaktovik. Did seismic testing cause the deep-sea fish to surface? With Shell poised to drill four exploratory wells in 2007, a federal court halted the project in response to a lawsuit by a coalition of Native Alaskans and conservation groups. Groups challenged the permit issued by the federal Minerals Management Service, on grounds that the agency failed to conduct proper assessment of environmental impacts. Stalled at least temporarily, the company will continue to press for development of its massive leases.

"It's just the start of what's going to happen," says Robert Thompson, an Inupiat wildlife guide. "Having the knowledge of what is planned and knowing that oil cannot be cleaned up in the Arctic Ocean is a foreboding feeling. If they have their way, we will have to contend with the possibility that a spill will affect the marine mammals in an adverse way to the detriment of our culture. We will have . . . this feeling for the rest of our lives and it will . . . pass on to our future generations."

In 2005, a majority of Kaktovik's voters signed a petition opposing oil development in the Arctic Refuge. In 2006, the village passed a resolution calling Shell "a hostile and dangerous force," and authorizing the mayor to take legal and other actions necessary to "defend the community." The resolution also calls on all North Slope communities to oppose Shell's offshore leases until the company becomes more respectful of the people. Kaktovik Mayor Sonsalla says Shell has failed to work with the villagers on how the company would protect bowhead whales.

To the west, another Inupiat village is beginning to doubt the benefits of oil development. With the discovery of oil in Prudhoe Bay in 1967, and the subsequent passing of the Alaska Native Claims Settlement Act, twenty-seven Inupiat families left Barrow to lay claim to their ancestral hunting and fishing lands. Where the Colville River delta spreads out before flowing into the sea, they founded the village of Nuiqsut in 1973.

Now, however, encroaching industrialization threatens their subsistence culture. The Alpine oil field lies just eight miles from Nuiqsut. Developed in the late 1990s, and touted as state-of-the-art oil-field technology, it was originally to cover just one hundred acres. A growing network of satellite fields expands the footprint exponentially, with the likelihood that Alpine will serve as the gateway for future oil development throughout the region. A 600 percent increase in respiratory ailments among Nuiqsut residents over the past decade casts doubts about the safety of oil technology. Like Kaktovik, Nuiqsut provides an onshore hub to serve as a platform for offshore development.

Nearby Teshekpuk Lake, once considered for inclusion as a national wildlife refuge by Congress, is also threatened. Teshekpuk Lake and the surrounding area support the highest density of nesting waterfowl and shorebirds on the Arctic Slope. Up to 60,000 geese (including lesser snow geese, Canada geese, and white-fronted geese and up to twenty percent of the entire Pacific black brant population) molt in the area. This is the primary arctic habitat for threatened spectacled eider and Steller's eider. Two distinct caribou herds make use of the Arctic Slope: the 500,000-member Western Arctic caribou herd and the resident 45,000-member Teshekpuk caribou herd.

Increasingly, as their homeland is industrialized, sentiment against further development grows among the Inupiat. They realize that no amount of money is worth losing a culture.

Wander around Anchorage and you sense an ephemeral society. The raw, elemental nature of the landscape contrasts with the tawdry, new veneer of civilization in Alaska's largest city. The problem with Alaska is that it got rich too soon, populated by wave after wave of boomers seeking riches in America's farthest-north resource colony. Before the United States owned it, the Russians came, raped the Aleut women, enslaved the men, and killed most of the sea otters. Then came the whalers, who did much the same, followed by the gold prospectors. Fish and timber extraction merely paved the way for the Prudhoe Bay oil boom. We even managed to create an economic bonanza after the *Exxon Valdez* oil spill. The ruination of Prince William Sound following the oil spill actually increased the GNP, according to Al Gore, because it put so many people to work, attempting to clean up the oil. Most of the state's riches have been squandered, while Alaskans search endlessly for ways to stay on the government dole to avoid paying taxes. Alaska has never had to grow up and deal with finding real solutions to serious problems.

Sliver by sliver, we whittle away at the land, destroying what makes Alaska so unique. The changes occurring here seem far more profound than elsewhere in the United States, as we adapt to the landscape and the landscape is forced to adapt to our desires. Ironically, much of Alaska is public land, so it belongs to all Americans. Yet we approve massive clearcutting of ancient forests, the largest zinc mine in the world, roads through public lands to access mining claims, mismanagement of wildlife and natural resources, burgeoning industrial tourism that chokes the regional character of small southeast Alaska communities, the spreading footprint of Big Oil, and the invasion of fast-food franchises and mega-box stores. The fight to save the Arctic Refuge is merely a synecdoche for all of our conservation battles.

Within the context of biological and geological time, Alaska is in the early stages of evolution. Although we barely understand the complex, mutually sustaining interrelationships between life forms, we've exceeded our privileges.

Deep within our hearts lies an ancient memory of our kinship with the wild. Are we bold enough, and wise enough, to honor this kinship? It takes imagination and dedication to practice restraint. Do we have the will to treat our wild lands as more than mines, more than oil fields for a temporary energy fix, more than adventure playgrounds, and more than laboratories for science research? Do we have the wisdom to know to leave some places alone, forever?

Ultimately, we will be judged not by our technological advancement, but rather by the enlightenment we exhibit in our care for the wolf, the grizzly, and every other creature—and for the mystery of life and its path of evolution.

Burns Cooper

Ice Auguries

Our false summer is finally over.
The storm in the tropics has quit
pushing air to us. I've had
to give up on my late tomatoes.
Below the 8th-floor window students shuffle
through the courtyard, hands in pockets,
heads tucked. The fountain's dry
except for a thin layer of rusty ice.
We haven't heard geese or cranes overhead,
their desolate cries, for weeks. The tourists
also long gone, the RVs and buses
and corporate herders. The birches
retain a few shriveled leaves
like Shakespeare's bare ruined choirs.

In my thirties, it feels like
the Thirties. All over now,
men with sandwich boards—
40, 50, 60 years old, trudging all day
looking down, shivering, silently hawking
furniture, cars, politicians, restaurants.
They're worth more now as signposts.
Their eyes are opaque.

Youngish men with hard little eyes and leather
jackets keep getting elected. They stand
for reaping what they have not sown, for raping
what they cannot own. How weary, stale, flat, and
unprofitable seems to them all the world
that they aren't using! Forests, ungulates,
marriages, exist to reproduce

and be harvested for Mammon;
wolves, foreigners, artists exist
for no good reason.

The TV predicts an El Niño winter—
our blanket of snow will be threadbare
and our summer dry. And yet
we still believe the geese will return,
many trees will survive, more of us will keep
our jobs than lose them. We will find
the spring sunlight glorious, even if
what it shines on is shabbier than we remember.

Martin Robards

The Mighty Sand Lance

That which we foolishly call vastness, is rightly considered not more wonderful, not more impressive, than that which we insolently call littleness.
—Ruskin

GO BACK TO YOUR POSTCARDS, videos, and magazines. Look at those puffins: Atlantic, tufted, horned, or even the rhinoceros auklet, sitting on ledges or swimming with a neat row of little fish lining that familiar, oversized bill. Look at the silvery specks flooding from the vacuous, baleen maws of hunting leviathans, or those fish spilling from the mouths of freshly landed halibut. Reduced in stature but not importance by their generic name of "forage" or "bait," these fish are the unsung heroes of our oceans. Maybe you are looking at the better-known herring or capelin, but often as not you'll be viewing the mysterious sand lance, a sleek spear-shaped fish with a steely blue back and silver belly, rarely exceeding much more than eight inches and about as thick as a pinky finger. This fish draws me out in rain, fog, snow, and sleet; in sun, moonlight, or clouded darkness; in wind and calm; wearing T-shirts or the whole wardrobe, onto the ocean and shores of coastal Alaska.

On this day, our small inflatable boat creeps through the water of Kachemak Bay, easing past the aptly named Gull Island. The conical outline of volcanic Mt. Augustine punctuates the western horizon, which is distorted vertically in the heat of this June day into incongruous columnar cliffs. It's easy to feel small amongst the immense views and glaciated mountains. Closer at hand, a three-dimensional, multisensory collage of life surrounds my research crew. Anyone who naively craves the solitude and freshness of nature shouldn't choose to quench that desire at a seabird colony. The mayhem and intense cacophony of thousands of screeching seabirds in the air, on the water, and perched on the island drowns out all other thought, and the omnipresent pungently sweet aroma of layer upon layer of guano, although reassuringly familiar to me, is likely to make tourists gag. Seals and a multitude of birds crease the water with their swirling wakes as they dive, bob up, and dive again. This is a scene that people who love seabirds spend their hard-earned dollars to see, and those who believe Hitchcock may have nightmares about.

The engine's gentle hum is lost to the intent gaze of our four pairs of eyes on the shimmering ocean surface. We try to peer through that barrier into more secretive and seductive realms, and see the culprit behind the commotion. We hope to glimpse schools of sand lance, fish that many have seen, but few have known. Throughout the coastal regions of the northern hemisphere, from the Arctic to the Mediterranean, this fish is generally viewed in its supporting role to the charismatic megafauna that draw tourists, fishermen, writers, and photographers.

The acclaimed "Father of Ichthyology," Swedish-born Peter Artedi, named the genus *Ammodytes*—literally, sand divers—in the early eighteenth century. It is ironic that Artedi, who dedicated his tragically short thirty years of life to fish, died underwater with the fish he loved. Friend and colleague Carl Linnaeus eulogized a copy of Artedi's magnum opus, *Ichthyologia*:

> Here lies poor Artedi, in foreign lands pyx'd
> Not a man nor a fish, but something betwixt,
> Not a man, for his life among fishes he past,
> Not a fish, for he perished by water at last.

By the end of the twentieth century, scientists had written over seventeen hundred articles describing sand lance and their part in the oceans. Their debut in more literary circles was in Beston's evocative classic of nature writing, *The Outermost House*, where they were immortalized along a 1920s Cape Cod beach:

> I saw that the beach close along the breakers, as far as the eye would reach, was curiously atwinkle in the moonlight with the convulsive dance of myriads of tiny fish . . . . I picked a dancer out of the slide and held him up to the moon. It was the familiar sand eel or sand launce.

Coastal communities around the Northern Hemisphere have all acknowledged this ubiquitous fish with its own name, from mainland Europe's *tobis*, *smeelte*, *equille*, *sandspierling*, *dobijak*, *agulhae*, and *lanzon*; to Britain's *sandeel* and *corr*; to the Faerose *nebbasild*; the Scandinavian *tuulenkalat* and *Småsil*; the Greenlandic *putorutotoq*; the Icelandic *sandsili*; the Baltic State's *peschanka*; and North America's launce, sand lance, and needle fish. Many names for a small fish, but the individual recognition by all coastal societies is befitting for a fish so integral to the ocean that surrounds it.

The six species of *Ammodytes* live between three and twelve years, if given a chance. Living on the cusp of land and sea in shallow waters, rarely deeper than one hundred meters, they are vulnerable in both environments. At least forty species of bird, twelve species of mammal, forty-five species of fish, and a few invertebrates such as squid favor sand lance as food, a multitude that pounces, plunges, and darts from land, air, and sea to snap, gobble, and suck them up in passing. Therefore, when not feeding or spawning, the sand divers secret themselves away within the seabed, their holy refuge and sanctuary.

Sand lance are visual feeders, needing light to feed, so most move out of their refuges to feed during the day on microscopic invertebrates. Those that survive the day's onslaught by hungry predators return to the same area of sandy refuge each evening. During the bleak winter months when their planktonic prey is sparse, they starve for long periods, remaining in their relatively safe, but not impregnable, blanket of sand, where they are able to survive for up to twenty-four weeks without food.

Despite undeserved anonymity in the public eye—there is no Disney *Sand Lance* or even a dedicated movie—the United States Navy recognized their sleek, streamlined form and potential for survival by emblazoning Sand Lance on the side of submarine SS-381 in 1943. Constantly under attack from the air and ocean, this vessel withstood one hundred depth charges in a sixteen-hour period—a true survivor, like its natural namesake, that will never receive five battle stars and Presidential Unit Citation for heroism.

As we make another pass through the feeding flocks at Gull Island, the effervescent fizz of glacial flour passing over our keel is soothing in a way that no Alka-Seltzer could ever be. My assistant Kali, leaning over the side, hair dragging in the water, exclaims that she sees a school. Hundreds, if not thousands, of sand lance hurtle around in a surreal nebula of movement, packed into a space barely bigger than a beach ball, expanding and contracting, as if a living entity in itself. Each fish intuitively closes the gap with its neighbors—safety in numbers, a classic expression of the "selfish herd" theory. Lithe, muscular bodies sparkle in iridescent sheens that range from silver through metallic blues and greens, depending on aspect and how the sunlight touches them. If these fish were larger, surely they'd be revered like the mighty and sleek tuna.

Unlike tuna, sand lance aren't generally eaten by people, although I did notice a bottle of salted sand lance sauce in an Asian market. However, they do make a tasty snack when rolled in spices and flour before a quick dip in the deep fryer; "french fries with eyes" was an apt description from a colleague in Homer. Lifelong rural Alaska residents Mike and Phyllis from nearby Seldovia also use them to fertilize their garden. But local use pales in comparison to the commercial sand lance fisheries of Japan and the North Sea, where they are the largest single-species fishery. In the North Sea, one million metric tons are caught each year for animal feed, fish meal, and oil—in direct competition with the two hundred thousand metric tons removed by seabirds. North America has no fishery for sand lance, leaving them to nourish the sea and those that live in and around it. Researchers in southeast Alaska have found sand lance deep inside forests, carried there and inadvertently dropped by secretive marbled murrelets. Because Pacific sand lance don't command the focused attention lavished on Alaskan walleye pollock, one of the world's largest single-species fisheries, their Pacific lifestyle remains largely mysterious.

The cacophony of screeching voices and splashing appears unending. It seems that all life in the area is in this tide rip, sand lance and those here to consume them. The attendees at the piscatorial banquet can be split into those above, on, or in the water. From above, raucous black-legged kittiwakes and glaucous gulls dive and stoop to the water. Paddling on the surface are the gaudy beaked puffin, gangly cormorant, and delicate pigeon guillemot with their Lady of Bath stockings. In the water are seals and the occasional flash of larger fish, including salmon, pollock, and flounder. Although they are not here today, sea lions, orcas, or an occasional baleen whale would be exciting but not unusual. Many gulls gather on the nearby shore like expectant children, waiting in line for the dinner gong as moon, earth, and sun align to produce some of the year's, and world's, largest tides. The ocean's level plunges twenty feet in just six hours, exposing wide swaths of intertidal hors d'oeuvres, including burrowed sand lance.

In the ocean as much as in modern human culture, you are what you eat. For marine predators, however, junk food is low fat and health food is high fat. Juvenile walleye pollock are the marine nutritional equivalent of lettuce, whereas sand lance, herring, and capelin are full of fat, and thus rich and nutritious. Sand lance are neatly packaged spears of energy, and perfect for numerous marine predators and the young that they strive to raise. Apart from being full of energy, the value of sand lance is multiplied by the size of their school. Each individual is safer within the school, but the school itself is vulnerable to larger predators such as whales, which can engulf a whole banquet in a single gulp.

We approach a small lagoon whose entrance is so plagued with shoals and sandbars that it is not uncommon to find vessels taking unexpected time on the newly exposed land, their captains nervously waiting and somewhat embarrassed. I'm always intrigued by how many large vessels find themselves so inextricably stranded upon exquisitely detailed and charted shoals and rocks. From little fishing vessels to tour boats, and even the tanker juggernauts, nobody is immune when the ship runs out of ocean. Pulling into shore, we purposefully anchor the boat above water, and even before the gaudy orange and overtly governmental float suits are removed, the vessel is high and dry. Grabbing shovels, we move along the shore in search of sand lance.

As a marine biologist, I am an evolutionary neophyte at catching these wily fish. They are easy to miss in nets, either because they are not there or simply because they are fast, streaming out of the net or into the sand for temporary refuge. But once their refuge is exposed, all I have to do is find them in their sand burrows, for then they have nowhere left to turn. On a suitable beach at an extremely low tide, I can wait for more experienced species to indicate the sand lance's precise location. In Alaska, there might be the ingenious corvids such as the crow and raven, or eagles, foxes, and even brown bears, all standing on the beach raking the sand with talons, beaks, and claws. I've never figured out how they make their discoveries; however, their quick searches followed by a precise peck or claw and the reward of a fish indicate that they know much more than I. If one of these predators doesn't find the sand lance, I might see a squirming fish rise into the divot left by my heel, then disappear in the next moment, back into the sand. Today I follow what are likely beak marks left by a crow, until, with a quick dig of my shovel and a well-aimed grab, I capture a sleek fish in my hand.

Holding the smooth six-inch sand lance, I can't help but admire this battle-scarred individual. Along the folded silver skin on its side (technically plicae) are V-shaped scars, most likely left by a disappointed bird. This fish, probably three or four years old, has survived constant hardship and threat. Now in June, it is already starting to assimilate bodily reserves into reproductive materials for the fall's spawning. The pouting lower jaw juts forward, a clever evolutionary adaptation that allows it to dig quickly into the seabed. The toothless, alienesque protractile mouth is also specially adapted for a life preying on small invertebrates. Unseen are other adaptations, such as lack of a swim bladder, which aids sand lance in a bottom-dwelling lifestyle, although forcing them, like sharks, to swim constantly while in the water column to prevent sinking. Large eyes precede a long dorsal fin that extends almost all the way to a forked tail. The features of this fish are all exquisite in their form, function, and beauty.

Without warning, the quiescent fish, in David and Goliath fashion, flexes, twists, and springs from my hand, disappearing into the sand as if a trap door had opened for it. Admiring the tenacious vitality, I tread gingerly and respectfully around the small impression, letting this survivor return home, to live and swim for some time more. I'll have to find others to measure and weigh as I attempt to understand their lives. Above the tide for at least an hour already, this sand lance could survive another four, easily long enough for the tide to return. Like many intertidal fish, sand lance have found ways to survive exposure, although we still don't know exactly how.

Even when the tide encroaches, or on the shallow, offshore North Atlantic Banks, such as Georges and Stellwagen, burial doesn't always save sand lance from the hungry mouths and unforgiving northern environment. Humpback whales that have discovered their habits just scoop up the sand along with the doomed, which are easily strained clean by sprawling plates of baleen.

I think back to a November night in Homer on Bishop's Beach when the mercury plunged below freezing and my rubber boots felt like chilled steel; at low tide I could hear the crisp cracks across the beach as the newly exposed surface, despite the salt, veiled over in an icy frosting. That was a death sentence for buried sand lance—forced to the surface where they froze like little gravestones, elongated shadows ominous in the beam of my flashlight. Likewise, in the extended exposure of today's tides and unusually hot temperature, sand lance appear around me, springing from the drying and rapidly warming sand, to wither and die like unwatered plants, or succumb to eagerly awaiting crows and kittiwakes.

Living on the junction between land, sea, and air makes sand lance available to more hungry creatures than may seem fair. It also places them in the path of all that is spilled and dumped, and disturbances that can render their precious sand refuges unusable. Although these steely lances that live life on the edge of everything are the embodiment of "littleness," we need to care about them and their part in the vastness of the oceans. They are the real deal, sustaining everything from the mighty leviathan to murrelets deep in coastal rain forests to farmer's fields in Europe and Japan. They epitomize everything that is important because, as a keystone species, they cannot be replaced. So, look at those postcards, videos, or magazines again. Take a moment. Shift your perspective from the charismatic predators to all that is mysterious and small. It is because of them that we have the images at all.

Jerah Chadwick

Morels

Spreading my sight
to scan the tundra,
I stumble, trying to
notice nothing
but the barest
bulb shapes. Rarest
fruit of spores
brought with wartime
lumber, they distill
overnight—leached
browns of leaf rot
ripening, a live
holding pattern
of decay. Green laces
the mosses, the bleached
grass sodden and thinned,
tendrils and shoots
edging soil rutted over
forty years ago.
Finding morels among them
only when I forget
to focus, wizened
hosts, I mouth
and taste one's
wild leavening, savor
its slow dissolve.

Bill Sherwonit

Out of the Depths

THE FRESHLY CUT HERRING BAIT has been sitting on Cook Inlet's sandy bottom for only a few minutes when there's a sharp tug on my line. My first impulse is to set the hook, hard. Instead, I follow Mark Chihuly's instructions: I wait. "Don't be impatient," Mark has been telling us all afternoon. "Let the fish take the bait." It's the same advice my Uncle Peach gave me when we fished live minnows for bass and pickerel in Connecticut lakes more than thirty years ago. Patience usually paid off then, and it's worked so far today: already I've landed a seventy-pound halibut, the largest fish I've ever caught and the biggest our party has taken on this splendid mid-July afternoon of blue skies, sixty-degree temperatures, and calm seas.

Waiting for the right moment to strike, I wonder what it would be like to catch a fish weighing two or three hundred pounds. I've read stories, even written stories, about people who've caught such halibut, but it's not something I've fantasized. Even today's seventy-pounder seems huge; the idea of hauling in something three or four times larger is impossible to imagine. Yet this spot, we've been assured, sometimes yields giant halibut. How would I respond to such a monster fish?

Other, more disturbing, questions have drifted through my consciousness today, like fishing line through water. It's not the first time I've wondered about my decades-long desire to hook and "play" hard-fighting sport fish. But as usual, I've refused to take that particular bait while caught up in the excitement of the chase and catch. Instead of insight, action.

Guided by Mark, five of us have been saltwater fishing from his charter boat, the *Suzy Q*, since midmorning, first for king salmon, then for halibut, a member of the flatfish family. The king fishing has been slower than slow; no one in our boat has even had a strike. "That's fishing," shrugs Mark, a stocky, muscular, good-natured outdoorsman and lifelong Alaskan who's spent much of his adult life guiding anglers and hunters. A favorite phrase, he'll repeat it a dozen times before day's end, as if to remind us—or himself—of the sport's vagaries. Still, I am disappointed. Allen, my brother-in-law, is visiting from Florida, and his best chance of catching an Alaskan king is past. Like Mark, I believe that "going fishing" means more than simply catching fish, but on this day Allen and I have come to the inlet expecting to take home both sporting memories and meat. As we headed away from the Kenai Peninsula coastline, I only hoped that halibut would salvage our day, make it memorable.

We anchor up miles from land, at a fishing hole where halibut of one hundred pounds or more are known to prowl the bottom. Back in the early 1980s, Mark tells us, it was not uncommon to catch exceptionally large halibut within a half mile of the beach. One year he pulled in a 325-pound fish and another weighing 250 pounds, "almost within a stone's throw of shore." But those days, he laments, are gone forever, ended by a saltwater sport fishery and charter industry that have grown precipitously in the past two decades. Trophy-sized halibut are still being caught, but anglers and guides have to travel farther and expend more time and energy to catch them. Each summer hundreds of charter boats work the central waters of Cook Inlet, a long, narrow, glacially fed estuary that stretches nearly three hundred miles from the Gulf of Alaska to the Mat-Su Valley north of Anchorage. And every one is after big fish: one hundred, two hundred, three hundred pounds. Any halibut that big is bound to be a breeding female. And the bigger the fish, the more eggs it produces: a fifty-pound female will, on average, release 500,000 eggs annually; a 250-pounder will produce four million.

Within minutes Mark has baited our hooks with herring and fish heads, and helped us to put our lines into the 160-foot-deep water. Joe, from Fairbanks, and I have fished for halibut several times before and know the routine. But Allen, Floria, and Kim—the latter two also from Fairbanks—have little or no halibut experience, so Mark re-emphasizes the two key elements to catching them: Number one, keep your bait on the bottom. Number two, don't set the hook too soon; let them really take it. Be patient, in other words.

Almost immediately, we get bites. Our first two fish are gray cod. Then a halibut takes my bait. I wait and wait, then set the hook hard. And again. The fish takes off, stripping a one-hundred-pound line from my bait-casting reel. I try to slow it by placing my thumb on the line, a foolish decision: it slices an inch-long gash into my skin. After its initial run, the halibut comes in slowly. Within fifteen minutes it's shot, gaffed, and pulled aboard. It's a good-sized fish and someone offers to take a picture. Grunting, I lift it off the deck. I'm satisfied. Now if only Allen will catch a nice-sized halibut.

For the next two hours, Mark keeps busy untangling lines, baiting hooks, unhooking fish. By three o'clock our boat has landed more than a dozen halibut and kept about half, most in the twenty-to-fifty-pound range; Allen pulled in one of forty pounds, inspiring big grins and a high-five exchange. Not big fish, but large enough to satisfy clients who've spent $150 apiece for the charter —and good eating. Small halibut are juicier, tastier, and more tender than larger ones. Mark, who clearly knows his fish, rarely brings home anything over twenty pounds for eating. Still, it's hard to shake the sportfishing maxim that "bigger is better." Given a choice, it's unlikely any of Mark's clients would trade a mammoth halibut for a small but tasty one. I've come away from charters disappointed that I caught fish weighing "only" fifteen to twenty-five pounds. Partly it's ego, but it's also the economics of guided fishing and a desire to fill freezers with halibut steaks.

Still looking for my second halibut, I feel a tug. I wait, then set the hook. At first I think it's a cod, but it proves to be a small halibut. With the seventy-pounder already boated, I'd be satisfied with this one to fill out my limit, but Mark tosses it back. Almost as an afterthought, he looks at me and says, "That was OK, I hope." It's as if he senses there's a bigger fish down there, ready to take the bait.

Again I let out my line. No more than five or ten minutes go by, when I feel the tug. It's a solid take, a steady, strong pull. I let the rod tip go down, let the halibut swallow the herring deeply. Then, finally, I set the hook: once, twice, three times, to be sure.

This fish doesn't run like my first halibut, but I can feel its heaviness. "I've got something," I say, feigning nonchalance, then begin a slow retrieve: lift, reel down; lift, reel down. This goes on for several minutes, until the fish decides to sit. I can't budge it an inch.

"Keep lifting," says Mark. "Stay with it."

I'm sweating now, grimacing. I'm surprised and a little chagrined that I can't move the fish. Where's that upper-body strength? On the boat's opposite side, Floria's line is doing funny things and Mark figures he's somehow become tangled with my fish. He tells Floria to give slack, then joins me. We work together, me lifting and reeling, Mark pulling the line, hand over hand, his heavily muscled arms and shoulders making my job easier. After five minutes of this, he looks at me and smiles: "You've got a big fish, maybe bigger than we've caught all year. We don't want to lose this one."

Everyone's attention is now focused on the fish at the end of my line. The others offer encouragement, tell me to keep at it. I'm lifting and grunting, working hard, but I'm also smiling, engaged by the growing drama. Slowly, reluctantly, the great fish is dragged upward until finally it is close enough to see through the murky water.

"It's huge!" someone shouts.

Mark peers over the gunnel. "This is the kind of fish that will get you in the newspaper," he says.

It's slack tide now, and the flat, diamond-shaped fish hangs from my line like a vertical slab of meat. Later, Mark tells me that most sport-caught halibut come up from the depths with their bodies in a more or less horizontal position. Doing that, he says, they actually help the angler, make it easier to reel them in. But occasionally a fish will come up vertically. When they do that, it's like hauling dead weight through the water.

There's another problem: halibut are powerful fish. Once boated, they're capable of doing serious damage with their flopping tails. In August 1973, the *Juneau Empire* reported the story of an Alaskan commercial fisherman who'd been killed by a halibut he'd caught: "The body of Joseph T. Cash, 67, of Petersburg was found lashed to the winch of his troller after a 150-pound halibut had apparently broken his leg and severed an artery when he hoisted it aboard his boat while fishing alone."

Because of this and other tail-thrashing incidents, sportfishing guides routinely shoot large halibut before bringing them aboard. Mark uses a.22-caliber pistol. The best place to shoot, he says, is in the spine, at the base of the skull. But when a jumbo halibut is hanging vertically, it's almost impossible to put a bullet in the spine. How do you lift a fish that size out of the water for a clean shot?

Slowly working the halibut toward the boat, we notice that another line is tangled with mine. Complicating things even more is the fact that my line has begun to fray. Mark fires his pistol. But instead of the expected fury, the halibut just sits there in the water. Now it's time to spear the fish, using a small harpoon that's attached to a rope and buoy; that way, if my line breaks, we'll still have the fish. When Mark strikes, the halibut thrashes wildly, then dives deep—without the harpoon, which somehow pops free.

"Oh no," Mark moans. "Give it line." That's what I'm already doing, not that I have much choice. The fish disappears into the inlet's grayish-brown depths and we begin the process again.

The rod butt is bruising my inner thigh, so Mark digs out a harness. But we can't get it to fit properly and I keep the butt pinned against my upper leg and groin area, trying to leverage the rod as I work it up and down. Tomorrow I'll have deep purplish bruises to remind me of this battle.

With the halibut wounded by the bullet, perhaps mortally, landing it becomes a more serious task. As Mark says, "It would be a shame to lose it now."

Praying that the frayed line will hold, I haul it up as gingerly as possible, inch by inch. Arms, shoulders, and thighs aching, I again bring the fish within harpooning distance and Mark jabs it. Again the fish dives deep. And again the harpoon point somehow breaks free, leaving a weakening line as my only link to the fish.

I sense a frenzy around me. Mark's nerves, like the line, are fraying: he curses our bad luck and scrambles around the deck; his clients scramble to stay out of the way. At one point Allen comes over to pat me on the back and offer encouragement. I'm a point of calm within the growing storm; hopeful, but without expectations or anxieties. I'm exhausted, but focused on my role, which at this point means simply holding on to the fish.

We bring the halibut in a third time. Mark decides to shoot it again, before attempting to harpoon it. I winch the fish until its upper body is above the water, amazed that the shredded line continues to hold. Mark shoots once, twice, three times, four times. I wince with each shot and my ears ring with exploding gunfire.

This wondrously large fish now hangs limply. Holding the line, Mark asks Allen to harpoon the halibut. He makes a quick, hard jab. "OK, it's in," says Allen, who looks my way and smiles.

The fish is ours now. Connected by the harpoon rope, it won't get away or sink from us even if the fishing line breaks. Allen sticks a large, hooked gaff into the halibut, then, with loud grunts, he and Mark haul it aboard. We shout our delight, exchange congratulatory handshakes. Only Mark has ever landed a halibut this big, and everyone marvels at the fish. Now that the fight is over, I'm dumbstruck. Exhausted. Relieved. How long has it been? A half hour? An hour? I feel as though I've been pulled outside ordinary time while connected to this giant fish by a thin thread.

I take some pictures then approach the halibut, which is lying on the boat's deck. The side that faces us is a brownish olive green, a predator's camouflage, while her underside, the side that faces the ocean bottom, is white. The skin is smooth and slippery—most would say slimy—when wet, but grows sticky with drying. I touch the skin of this great-grandmother fish and something shifts in me: for a short while, at least, "it" becomes a "she." I watch her gills slowly expand, remain open for several seconds, then collapse. Is this a reflex, I wonder, or does some life force still flicker within? Except for the gills, she's still.

I'm drawn to her large, bulbous eyes, which protrude from the green upper side of the head like golf ball–sized knobs. Larger than a nickel, each black pupil is surrounded by a golden iris halo. Though unfocused, the eyes have an eerie depth to them. They pull me in, enchant me.

"It's a beautiful fish," says Allen, as though reading my mind.

"I'm glad you said that," replies Mark. "Most people think they're ugly, but I've always found them to be pretty fish."

Their comments make me consider how we so often label these large fish "monsters." So-and-so landed a monster king or a monster halibut; I used that phrase many times when doing fishing reports as a newspaper writer. Later, a female friend will point out that such monster imagery seems a very masculine thing. Perhaps it adds drama to the hunt, accentuates the idea of doing battle with a large and powerful adversary. The paradox, in this case, is that such monster halibut are females. More than that, they're fecund mothers. I'm reminded that Western man's traditional attitudes toward hunting have sometimes been linked with their passion for sexual conquest. Both beasts and women become objects of desire, something to chase and conquer. So which is it: beauty or beast? A little bit of both, probably, in halibut as well as humans. And where do I fit in as fisher, as hunter? Surely while "fighting" the halibut, I felt something close to desire. Now other emotions begin to intrude.

As others resume fishing, I fall into a cushioned seat, physically and emotionally drained. Earlier I'd been fully absorbed by the chase, the catch. Now I wonder if I should have given the rod and the experience to Allen, who in recent years has become an avid, if still largely inexperienced, fisherman. Traveling north with his wife and three daughters, Allen has enthusiastically spent most of his Alaska vacation on sightseeing and camping trips to Denali, Resurrection Bay, and Kenai Fjords. We've caught a few small fish along the way, but this is the one day he and I have devoted solely to angling. What an Alaska memory this great halibut could have been for the middle-aged businessman from Florida. But while I was hooked to the halibut, the thought of handing the rod to Allen didn't even cross my mind. Nothing in his actions or words suggests envy. On the contrary, he seems as excited as anyone about my fish. Later, heading to land with his modest catch, Allen will voice satisfaction in his own harvest. Still I wonder: was I selfish?

I wonder too if killing this halibut was the right thing to do. Based on body measurements, a biologist later estimates her to be twenty to twenty-five years old. Something about her age, her size, her eyes, combined with what Mark has told me about the fishing pressure on female breeders, make me question my own desires and attitudes. Perhaps it's the knowledge that, during our struggle, I was caught up in the battle and gave little thought to her life. Or death. She was something to be caught, a form of wild game—a slab of meat. I roll the words around in my head: something to be caught, some *thing*. Before seeing her up close, the halibut had been an "it," a resource, not another being.

I think back three days, when Allen's young daughter, Emily, caught a sea-run Dolly Varden near Seward. Excited by the catch and charmed by the char's sleek beauty, Emily wanted to keep the fish. She would show it to her family, then we'd have it for dinner. I grabbed a rock and clubbed the Dolly, but not well enough to kill it immediately. The bloodied fish flopped on the beach. As I prepared to club it again, Emily had second thoughts: "Maybe we should let it go," she said softly. "It's too late," I told her gently. "It wouldn't survive."

Watching her stare quietly at the dying fish, I asked what she was thinking, feeling. She only shrugged, a half smile, half grimace on her face. Still silent, she bent and touched the fish, then slowly picked its limp body from the beach and carried it to her father. Later, Emily and I talked some about the fish's death, and how it was natural to feel such contradictory emotions: the initial rush of excitement when catching and seeing the fish, then sadness, even regret, at the taking of another creature's life. Killing is not an act to be taken lightly, I reminded her. When necessary, as for eating, it should be done with respect for the animal's life.

Now I'm having the same sort of second thoughts. It was only after I'd seen the old halibut's final breaths, looked into her eyes, and felt her flesh that I'd fully comprehended what I'd done: taken a life. It's strange how the earlier, smaller fish we'd caught didn't affect me this way. No brooding with them, no self-analysis. Even after considerable reflection, I can't explain the difference. Maybe it's because the smaller fish were so quickly dumped into the holding tank, leaving no time for connection or remorse. Maybe it took something as huge, as miraculous, as this great mother halibut to shake things up inside.

Could I have seen, somehow perceived, her life flickering away? Did the halibut speak to me in some way? I want to honor this matriarch somehow, to show my respect for her being, her kind. Sitting in my chair, I look again at the halibut's unblinking eyes, her huge and slightly opened mouth, with its rows of sharp teeth. In her world, she's at the top of the food chain, an aggressive bottom-feeding predator who eats crab, shrimp, schooling fish, cod, even other halibut. Now she's become the prey.

There's something else about the catch that bothers me: the "sport" aspect. We killed this halibut—all these halibut—for meat, but also for sport. For the thrill of the chase, the battle, the catch. Tutored by my Uncle Peach, a passionate angler who never showed guilt or second-guessed his motives, I've been a sportfisherman since I was eleven or twelve. Sometimes gruff, often playful, Peach was (and is) a boisterous, bearded Hemingwayesque character with a passion for the outdoors. Under his guidance I learned to catch night crawlers and minnows for bait and place them on my hook. I learned patience and the proper way to present the bait, to set the hook and play the fish. We were fishing buddies for years, sometimes filling our limit, more often getting skunked. Almost always we kept what we caught in Connecticut lakes and streams, the fish becoming food for the table. But sport, not food, seemed to me our main motivation.

Later, when I moved to Alaska and took up fly-fishing, I became an avid catch-and-release angler. In the world of fly fishers, catch-and-release is the highest, most ethical, and conservation-minded form of fishing possible. Anglers, biologists, and conservationists agree it has protected many fish populations that would otherwise be decimated by heavy fishing pressure.

There still are times when I enthusiastically embrace such sportfishing. But at other times—like now, on this halibut charter—I have second thoughts. The idea of harming another creature for sport, as opposed to need, is disgusting somehow. We have sport or game fish and we have big and small game animals. What it becomes is blood sport. A bloody game.

The questions go beyond simple killing. Nearly every sportfisherman and sport-fish biologist I've ever met has accepted as truth the notion that fish feel no pain. At least not the way we humans do. Or even mammals and birds. That makes it easier to hook 'em, haul 'em in, and then either kill or release the fish "unharmed." I wondered about this even as a boy. But whenever I raised the question, Uncle Peach always assured me, "Fish don't feel pain." And that was that. But is it true? Or are we humans torturing and sometimes killing these critters for fun? For play?

Because they are cold-blooded, live in water, have fins and scales instead of limbs and hair (or fur or feathers), we don't identify with fish the way we do with our warm-blooded relatives, mammals and birds. It becomes easy to rationalize that they don't have feelings, can't experience trauma—or if they do, it's no big deal. As philosopher (and former angler) Jack Turner has commented,

> Imagine using worms and flies to catch mountain bluebirds or pine grosbeaks or maybe eagles and ospreys, and hauling them around on fifty feet of line while they tried to get away. Then when you landed them, you'd release them. No one would tolerate that sort of thing with birds. But we will for fish because they're underwater and out of sight.

Yet for all of our rationalizing—and denial—there's increased scientific evidence that fish do in fact feel pain. One researcher, Michael Stoskopf, found convincing evidence that fish (and other nonmammals) had both biochemical and physical responses similar to those of mammals when subjected to pain. A growing number of other ethicists and scientists say it's unthinkable that fish don't suffer when hooked and pulled through water. Should this matter?

Tied to the pain of catch-and-release fishing is the question of respect for other life forms. Back in the mid-1980s, while working at the *Anchorage Times*, I reported on fishing conflicts along the Kanektok River in southwest Alaska. Native residents of Quinhagak, a Yup'ik village at the river's mouth, had been protesting the increased presence of sportfishing parties and their impacts on local subsistence uses of the Kanektok. I've forgotten many details of that conflict, but one thing has stayed with me. I remember that the villagers were especially offended and outraged by the visiting anglers' use of catch-and-release techniques. To locals, the practice was disrespectful to the fish. It was, in essence, playing with the food, a taboo in their culture.

Initially I found it strange that catch-and-release might be offensive, even "bad" or unethical. But after listening to the Native perspective and reporting on this clash of cultural values, I began to reassess my own beliefs and biases. The more I thought about it, the more the Yup'ik attitude made sense. Even the terms "sport" fish and "game" fish imply the notion of play. It doesn't take much of a leap in logic to see that this Western art form easily could be seen as playing with food, especially if the fish being "played" (as anglers like to say) is an important part of someone else's subsistence diet, a source of sustenance that's important for survival.

For all my growing doubts about the sport of fishing, it seems more honest to kill the fish and other animals we eat, rather than avoid or deny the loss of life by shopping in supermarkets for prepackaged products. So many in our culture—and I am certainly among them—have become distanced and disconnected from the food that gives us life. When done properly, fishing and hunting allow us humans to claim and perhaps even celebrate our role in the food web, our connection to the world that sustains us, instead of denying it.

Mark might respond, "That's fishing." At least for me such back-and-forth agonizing has apparently become a part of the ritual. A key to acceptance, I suspect, is to treat this halibut, or any animal, with respect; to have the humility to recognize this halibut as a gift, not a conquest. Many indigenous peoples believe that hunting success has less to do with any great skills than the fact that prey presents itself to the respectful hunter. Their traditions emphasize humility and restraint, a recognition of spirit and sacredness in all of nature, and a receptivity to mystery. More and more, such beliefs and mores make wonderful sense to me.

Perhaps picking up on my somber mood or my questions about the halibut's size and age, Mark says he'd like to see a limit placed on the number of hundred-pounders that a sportfisherman can keep. After one lunker, what's the need for more? He then recalls a story from the previous summer, about a fishing charter that brought in ten or twelve halibut weighing one hundred to three hundred pounds. Mark sees that not as some special triumph, but a waste. The big females are the ones that need protection, he again emphasizes. Why not have a system in which it's possible to arrange some sort of halibut trophy mount, without killing the fish? It's already done with certain species, like rainbow trout.

Fishery biologists who monitor Cook Inlet say the increased sport harvest of breeders isn't threatening the inlet's population. But their assurances don't ease Mark's worries. The inlet is home, it's where he has fished since childhood, and he sees a local problem developing if guides and anglers don't ease up on large female halibut. "Somewhere down the line, it's going to catch up with us," he frets. "We need to show some restraint."

But restraint is a difficult thing when you own a charter operation. Trophy fish make for good advertising. The walls, doors, and windows of charter offices are plastered with pictures of barn door–sized halibut. And at day's end, crowds naturally gather around the largest fish. "He who can get the biggest fish is he who will get the business," Mark says. "You'd be cutting your own throat if you didn't go after the big ones. But what's going to happen when they're gone?"

Mark then tells me about a halibut sport fish–tagging program run by the International Pacific Halibut Commission. Even as we're catching our limits, there's another Chihuly charter boat on which customers are catching and releasing large halibut: by day's end, several fish weighing from eighty to a hundred and fifty pounds will be tagged and let go. The tagging program gives people another option. Guides buy tags from the halibut commission and, if clients are interested, they tag their fish, get an estimated size, and turn them loose. In return, tagging participants receive a pin and certificate, and if a tagged fish is eventually re-caught, that information is passed on to the original halibut catcher. Tagged and recovered fish, in turn, give the commission helpful data on halibut movements and growth.

Recognizing that his tagging clients might wish to preserve some record of their released fish, Mark has constructed a half dozen plywood replicas of halibut. Each a different size, the plywood pieces are cut and painted to look just like the fish; they even have blood running out their gills. I'm not sure about getting my picture taken with a plywood halibut, but I like the tagging idea. Right now, I can't imagine myself ever again keeping a trophy-sized halibut; the idea of returning grandmother fish to their saltwater homes pleases me.

On our return to shore, Mark suggests I get a tail mount of my halibut. The fish's tail fin measures twenty-one inches from tip to tip. At first I'm repulsed by the idea: for years I've opposed the act of collecting animal trophies, whether fish or bears. Big-game trophy hunting especially bothers me. Killing for meat I can understand; but killing for heads or hide alone? Killing to place a prize upon a wall or floor? This deadly game reeks of self-aggrandizing pride and male ego, at the expense of another's life. I find it offensive, unacceptable.

Is a halibut tail mount any different from a grizzly bear mount? I eventually decide it is, at least partly because of intent: I didn't come here looking for trophies, I came to harvest meat. And yes, engage in sport. But there's something else: a tail mount would preserve the memory of all that's happened today, including the inner debate that has raged since I peered into the halibut's eyes, touched her skin. In a sense it could be my memorial to the great fish, a way of honoring her spirit. Placed on my wall beside a picture of the halibut, it will be a reminder of the day's ambiguities and revelations, and my own contradictory nature. This is not one in a series, but once in a lifetime. After still more agonizing—am I being hypocritical?—I decide to do it. For $125, a local taxidermy shop will prepare the halibut's tail fin, to be hung upon my dining room wall.

Back on land, we weigh the halibut at Chihuly Charters. The scale reads only 220 pounds, which surprises us all. Mark explains that it's uncertified and only gives approximate weights. Next we measure the fish: she's eighty-two inches long. The International Pacific Halibut Commission has put together a table that correlates halibut length and weight; by its measure, an average halibut of eighty-two inches would weigh 299.5 pounds. So I jokingly split the difference and figure my halibut goes about 298.

Chihuly's, like most charters, has a "hang 'em and shoot" ritual for its customers. Halibut are hung, white side out, on a fish rack, and their catchers are then posed with the fish. As we bring in our catch, mine gets everyone's attention. It's strange to be at the center of this commotion, both exhilarating and embarrassing. Soon our images will join others on Chihuly Charters' office door and the big halibut in our Polaroid will grab people's attention. Mark may not like it, but that's part of charter fishing too. My trophy catch may symbolize the growing pressures on Cook Inlet halibut, but it is also good for business.

After the photo session, Mark and his guides fillet the halibut. We catchers of the fish stand aside, curiously detached from this part of the harvesting process. Even after the head, guts, and bones are removed, Allen and I will take home more than 250 pounds of fish from our four halibut. I get caught up in the picture-taking frenzy, the meat packing, the preparation for our drive back to Anchorage, and the halibut, for a while at least, again becomes an object—something for the freezer.

On Sunday, the day after our fishing expedition, Allen and I cut the halibut fillets into smaller, meal-sized pieces and bag them for refrigeration, a three-hour process that becomes part of our shared ritual. He'll take a share back to Florida; I'll keep the rest here, giving some to friends. It's satisfying, to have a fresh supply of halibut for the winter. Last year I didn't put any fish in the freezer.

I barbecue several pounds of the fish for Sunday dinner and seven of us hold hands as Allen thanks God for the halibut. Silently, I add my thanks to the halibut for giving herself to us. Then we eat.

Peggy Shumaker

Walker Lake

The sow bear ripped
down the boat tarp,
scraped black fur

into twisted wing nuts,
that bear
eased her itch and disappeared.

And still she stayed near,
while we hiked
uphill to fill our jug,

the spring covered over
with autumn's leavings.
We skimmed clear

frosted growth floating,
sank our jerry cans,
felt them pull deeper.

The surface healed
around wrists
stiff with cold.

Bubbles shook free
from river weeds, rose up,
tumbled downstream.

The still place returned
to reflection. Birch
startled us with gold

so loud our bodies
flared like fireweed
gone to seed.

Twilight
put its slant
on the afternoon—

four loons
swam near
our seaplane,

nudging this
strange relative
who would not speak.

The moon had eaten itself
down to the rind,
and in that sliver,

that lingering
of autumn, stars borrowed
the voices of loons.

Nancy Lord

Fishing Grounds

HERE AND NOW: THE HARD RAIN FALLING, and the hills dim and gray with a yellowed, washed-away cast. The seiners and gillnetters anchored around us, riding out the weather within Ikatan Bay. A kittiwake chopping past. The anticipation.

Today is the first opening in Area M, and there is not a soul for hundreds of miles who doesn't know this, who isn't involved in some intimate way with the fishery that sustains this region and its families. Beyond salmon, beyond weather, beyond testing new gear and crew, there is nothing that matters. Four hundred boats fish here, and for everyone on them, and on the tenders and processors, there is no other world.

From the pilothouse I watch through blurred windows as our crew tears apart cardboard boxes and loads resale goods into our freezer and storage. Steaks, ice cream, bags of bread and jars of peanut butter—these, too, are part of our service to the fleet. Beside and above and behind me the multitude of radios booms and cackles. I am to listen for anyone calling us, for weather, for news of catches, prices, species ratios. The two VHFs are set to different channels: the single sideband can leap voices over mountains; the mysterious black box, on its private frequency, communicates solely with our company and its other tenders.

Fish buyers are calling through the channels, advertising prices, bidding each other up. $1.30. $1.35. $1.40. Cash. Bonus later. Everyone talks reds, no one wants to mention the "other" fish, chum salmon. The real news, we know from the company, is that too many chums are being caught. The chums are meant to get through for fisheries farther up the coast. Too many intercepted here, and the whole fishery will shut down.

On the radios, fishermen growl among themselves in protective code. Their locations are vague, their catches shared in senseless syllables, in phrases such as "coconut pie' and "betty to veronica," in Russian. We, too, will be reporting numbers of boats and pounds in a scrambled code provided by the company office. "Zulu boats." "Hotel alpha foxtrot thousands of pounds."

From the sheltered bay, we're sent back out to the fishing grounds. Ken navigates with binoculars held to his eyes; there are boats everywhere, and it can be hard to see and avoid their nets. Few boats, though, are actually fishing at the moment. They're running from one place to another, or they're waiting—for calmer seas or more fish, preferably both.

In East Anchor Cove, the anchored boats all lie with their bows into the wind, like birds of a flock. We easily locate our friends on the *Lucky Dove*. Buck and Shelly and their two crew members are in from tending their nets off the south side of the peninsula, where they had all gotten sick in the rocking and rolling.

A few pints of Häagen-Dazs from our freezer, and they are revived.

Later, when the weather calms, we motor out to their sites to collect fish. It is an exposed, rocky shoreline they fish along, with the sea crashing in and long, gray, lost-in-the-clouds hills behind. The *Lucky Dove* serves as shelter for eating and sleeping, but their fishing—set gillnetting—takes place from small open skiffs that can work close to shore, the fixed nets drawn across their gunnels. We watch from our anchorage as the nearer skiff tosses in the waves, the orange-hooded figures bent over the net, pulling and picking, snapping salmon from the web: a very small tableau, pinpricks of color, against a dramatically large and somber landscape.

This is a romantic vision, yes, that we cling to, men and women working the sea in this time-honored fashion. Not many do it like this anymore, one fish at a time. It is not the most efficient way, certainly, nor the easiest or safest. But the salmon fisherman knows an art, and he knows about weather and whales and plankton blooms—all those connecting things that make up the salmon's world, and his own.

The writer John Burroughs said, "Knowledge is only half the task. The other half is love." He was talking about writing natural history, about knowing facts and not stopping with them, but he might as well have been talking about fishing.

North and south of here—in the Bering Sea and in the Gulf of Alaska—modern, industrialized fishing ships are also at work. Giant factory trawlers, dragging enormous nets that can catch four hundred tons of fish in a single tow, fish not for salmon but for groundfish like pollock and cod. They tear up the sea bottom, they waste enormous amounts of fish of the wrong sex, size, or species (including salmon) that they catch accidentally, and when they have fished out one area they move on to another. Little understanding beyond mere technical knowledge is involved, and, I think, no love at all.

My friends Shelly and Buck have noticed that when the pollock season opens in the Bering Sea, unusual numbers of Steller sea lions pass through the strait beside their home, heading south, away from the fishery. The animals, now listed as an endangered species, appear to be either fleeing the disruption caused by the fleet of factory trawlers sweeping the area or searching for food to replace the pollock removed so precipitously by the fleet. It is a fact that sea lions and other sea-feeding animals have fared poorly in recent years in areas of heavy trawling. It is another fact that today's scientists—specialists all—are reluctant to draw any conclusions. Each expert studies his own little piece of the ever more complicated ocean environment, and no one ever knows enough.

This is my sober thought: we have come so far from the time when someone like William Dall, Alaska's first scientist, could excel not only in a broad range of the sciences but also in history, geography, anthropology, writing, and understanding. Our gain in specialization is also our loss, until perhaps only small-scale fishermen and their kind are left as the generalists who see things whole—and who will defend not their disciplines but our lives.

Joan Kane

Withdraw

To the south of those who live south of us—
I will visit unknown men, hunt up the invisible
Behind the women.

It is both a bluster and a promise,
The day immature, blunt in its newness.
I do not know what steals me

Above the spines of mountains.
I trespass to hear the sound of the sea,
Its resemblance to a summit of wind.

Child, I pare off. The swallows
Have disappeared into their banks
And emerged as wolves. Expect

A bird of beak and tooth, the steep
Fetch, the sigh of new-formed ice.

Nick Jans

Wolf Wars

MY WIFE, SHERRIE, and I stood in the foyer outside the Nome post office, clipboards in hand. Out on Front Street, snow spattered on a blustery west wind. I'd just come from three days in Kotzebue, where I'd spent hours at a time outside at fifteen below, rotating a half dozen pens from an inner pocket, trying to keep each from freezing long enough to scratch a signature. By comparison, this was Miami Beach.

A steady stream of people, all on missions that didn't include talking to us, bustled past. A grizzled gold-miner type in worn Carhartts held my eye and nodded politely—the sort of guy I'd have a beer with. And, for the umpteen-hundredth time in two days I nodded back, stepped forward, and said, "Excuse me, would you like to sign a petition to help stop the state's program of shooting wolves from planes?"

He stared back incredulously. "Stop it? Jesus, if I had a plane, I'd like to get a few of the bastards myself! They're eatin' all our moose!" I stood there, watching the greasy back of his jacket recede, feeling an idiot. Out of the corner of my eye, I saw Sherrie approach a grandmotherly Eskimo woman with an armload of packages. "Have you heard about our petition?" Meanwhile, a young man with a toddler in tow ambled by. I sighed to myself, and heard my voice, tired at the edges, once more. "Excuse me . . . " The guy stopped, smiled, and took the clipboard from my hand. Thanked me, in fact. One more down. No, two, with Sherrie's. Three hundred-something more to go in Nome, thirty-some thousand statewide.

Don't ask me how I got into doing something I loathe so much. I'm not talking about the cause, but about being political, and the idea of interrupting folks minding their own business and asking them to jump through some hoop. Bad enough in Juneau or Anchorage, but far worse in the bush, where the unwritten code is do-what-you're-doing-and-mind-your-own.

Not only was I gathering signatures. I'd somehow ended up co-sponsor of a statewide ballot initiative to limit aerial wolf control for the third time in a decade. Twice a majority of Alaskans had voted against the practice and banned it by law; and twice the governor-appointed Board of Game had reinstated the program as soon as a two-year statutory limit had expired, to be used as a management tool over broad areas. This time around, they were permitting private pilots to do the actual shooting.

It didn't take me long to reconfirm what I already knew: I might as well have signed up to sit on a lightning rod. Aerial wolf control has long been Alaska's most controversial wildlife management issue, the sort of topic that leads to hard feelings, finger-jabbing, nasty letters to the editor, and occasional bar fights.

Two opposing philosophies define the argument. (Ahem.) Ready?

Position A: Wolves constitute a looming predatory menace to the game animals on which the people of Alaska depend—not to mention a threat to human safety. Keeping their numbers under control by whatever means (including shooting, snaring, leg-hold trapping, and shotgunning them from low-flying aircraft) is a common-sense necessity. Left to their own devices, wolves will multiply and Hoover every moose and caribou out of the country. People come first, and Alaskans have a right and legal mandate to manage wildlife for their own maximum benefit. Any opposition to such a plan obviously comes from greenie-weenie, barely Alaskan, nonhunting city slickers and out-of-state radical, pinhead lackeys of PETA (People for the Ethical Treatment of Animals).

Position B: Wolves, as top predators, are a natural part of healthy, complex, self-regulating ecosystems that have evolved over millennia, and removing most of them (the plans call for up to eighty percent in certain management units) is only bound to screw things up. Without wolves, deer and moose populations explode in unsustainable numbers, then crash, over and over. Wolves, too, are a valued resource on which trappers and subsistence hunters depend. Beside that, blasting wolves from airplanes is just plain wrong and reflects horribly on the state's image. Anyone who doesn't see things that way is a nearsighted, beetle-browed, knuckle-dragging redneck.

That's just the Cliff Notes summary. The unabridged version gets far more nasty and multilayered, replete with biologists, politicians, wildlife advocates, and hunters flinging mud balls made of statistics and rhetoric in each others' faces. Add in the real extremists—old-schoolers who consider wolves four-legged cockroaches, and the animal-rights types who worship *Canis lupus* as imperiled uber-beings—and you have the makings of a full-scale brouhaha that spills over state and even international boundaries. Wolves, by virtue of their innate canine charisma and endangered status through most of their former range, are a big deal. People far away care what happens here—a fact that rankles many Alaskans, who believe wolf control is no one's business but their own.

Alaska's wolves are unique in at least one respect. At the dawn of the twenty-first century, there's still plenty of them—statewide, somewhere between seven and eleven thousand, according to state biologists. Thanks to the elusive, no-paparazzi nature of the species and the scale and roughness of the country, these are educated estimates at best, with a huge amount of slack (more than fifty percent the minimum figure) built in. Some biologists figure it's more like five to seven thousand. But whatever the number, some folks—especially those associated with the big-dollar sport hunting and guiding industry, who consider every bull moose a walking paycheck, and a few thousand rural residents living in relatively game-poor areas—figure it's too many.

I could run down the whole time line of Alaska wolf control, from federally sponsored bounties, government hunters, and cyanide-laced baits of the territorial days through a period of more enlightened, ecosystem-based wildlife management, to the current tug of scientific evidence and ideologies, but that's its own convoluted story. Somehow, though, all that led to us and dozens of others standing with clipboards all across the state, gathering signatures that would give voters a chance to reaffirm what they'd already decided twice: shooting wolves from the air wasn't Alaskan or right.

My own history with wolves isn't what you might expect. One of the reasons I headed for Alaska twenty-eight years ago was that wolves still roamed wild here, and I wanted to be part of that landscape. Naturally, I wanted to get close to them, interact somehow—which, to a twenty-something kid raised on *Outdoor Life*, meant hunting. Not in a systematic, specific way, but wolves along with everything else, from grizzlies to Dall sheep. I launched my education as a packer for a big-game guide, then honed my skills alongside the Inupiat hunters who were my friends and neighbors for twenty years. And with time, I got pretty good at hunting most things—enough so that after a few years I lost count of the wolf hides, even though I often passed up fresh trails and easy shots. The skins ended up as parka ruffs, decorations, and gifts to village elders. Meanwhile, I never saw any subsistence hunter who truly needed a moose or caribou go without.

I stopped hunting wolves as a matter of personal choice, mostly because, through long familiarity, I started liking them much better alive. An empty hide didn't have eyes that flashed yellow fire, didn't flow across the tundra with effortless, loose-wristed grace, or play with ravens and sticks, howl unseen from a ridge, lead wobbly pups past camp, taunt grizzlies, and sometimes cavort with my dogs. Pull the trigger and all that was gone, reduced to a bloody pile of hair and meat. Dangerous? Potentially, sure. But in dozens of meetings, sometimes as close as twenty feet, with healthy, wild, full-grown wolves from the North Slope to Southeast, I'd never had the least hint of trouble. Moose were one hell of a lot more risky.

Not that I had any illusions about what wolves were, what they could do, and how they lived. I'd seen dozens of kills over the years—moose, caribou, Dall sheep—some so fresh the gut piles were still steaming. Wild wolves struggled for dominance and often killed each other. They starved, died of mange, got their heads stove in by moose kicks, went days at below zero without eating, and got run down by hunters on snowmachines. The miracle was that they somehow managed to survive at all. And even when they were abundant, they were spread so thin over the land that most Alaskans have never glimpsed a wild wolf, or heard one howl. But they should have that opportunity, and not just in some national park. They and their grandchildren should be able to legally hunt and trap them if they want, too.

Wolves, even unseen, fill up a landscape with wildness, define it. You'd think, after the mess we made elsewhere, that people would know better, learn to value the last places where large-scale ecosystems without boundaries exist, complete with the predators that define and shape them. Am I overreacting? The Position-A guys would say way worse than that—I'm ranting against nothing, fear mongering, distorting, and promoting ballot-box biology. They just want to *manage* wolves, not eliminate them, and their science is good. There will always be wolves in Alaska. Couldn't get rid of them if they wanted to.

Sorry, guys, I'm not comforted. You don't remove eighty percent of a population of social, pack-oriented animals without getting rid of them all. And sure, just in some game management units for now, but the working plan is to expand predator control areas, not reduce them. Check the minutes of recent Alaska Board of Game meetings. Think we lack the technology and will to exterminate wolves? Compare a map of former with current wolf range worldwide and get back to me. As for science, there's plenty of respected biologists on the Position-B side specifically questioning that "good science" behind the state's program, poking holes in faulty data and methods, and pointing to issues of sustainability. I try to imagine big chunks of Alaska essentially wolfless and Pennsylvania-like as some would have it, and three things happen. First I get unspeakably sad. Then comes the anger. And then I head for the post office.

Anne Coray

Election Day: Reading Chevigny and Thinking About Referendum Six

A yes vote overturns the state legislature's recent reauthorization of same-day airborne land-and-shoot wolf hunting that an overwhelming majority of Alaskans voted to ban in 1996.
—State of Alaska Election Pamphlet

Someday I will write a poem wholly praise
when light perfects a quiet temple
over the wide, insistent village
and a glad sea wells in the trough of these bones.
Not today. I've finished another history
and know the *promyshlennik* soul that breeds
in the black, thick-blooded bowel.
We vote again on another measure of slaughter,
an old *shitik* put to sail.
It's not just outcome, but the iron bell
with its iron tongue: and we shall have dominion . . .

Outside, clouds are gathering. Soon it will snow.
There are moments I'd like to walk out,
find a little hill and lie down, and let the sky ferry
me into a land so clean it exacts no witness.

promyshlenniki: *Freelance exploiters of natural resources, notably fur.*
shitik: *The first vessel used in the Russian quest for fur.*

Douglas Yates

Partners on the Wheel: Dancing at the Vernal Equinox

IN THE NORTHERN HEMISPHERE, the sun's measure on March 21 marks the vernal equinox. Here in Alaska's Tanana River basin, the event is more than an illuminated contour on an upland ridge line. It's a place where travelers point out the territory that's been crossed while turning a hopeful eye to the landscape ahead.

Celestially, it's a quiet passage. A span of equal parts of day and night that lasts just twenty-four hours. But in the high latitudes the symmetry closes a cycle and opens a shared vantage. Six months ago, bidding farewell to summer's migrants—many birds and most tourists—we accepted the land's new austerity, snow settling over its scarps and rough edges.

With the sun's retreat and winter's imperatives, year-round residents are either indigenous or well equipped. In deep snow years, moose yard up in the big aspen on the hillside behind my cabin, stripping bark for food. Sheltered from the cold by superinsulation, hearth and home are defended with psychic armor while residents burn rounds of spruce and birch. Television sells diversion; beyond the window thermometers order reality.

In cold isolation, the dry crunch of snow underfoot is a steady companion, like a metronome, measuring the distance between the cabin and the woodpile. Feeding the stove is a ritual that warms twice, splitting and hauling wood first, then later as it's consumed by fire. Cold-soaked wood lends another advantage: in subzero temperatures, a splitting maul cleaves birch as if it were ice.

Following December's solstice, the new year welcomes a ridge of frigid high pressure. Pilots call it "severe clear." If it stalls overhead, temperatures plummet. By mid-February, with sunlight increasing at a rate of six minutes a day, the dazzling albedo puts me behind sunglasses and winter in the rear-view mirror. An old-timer in Ester once told me, "The first day of spring is when the squeak goes out of the snow." And none too soon, I've grown impatient with winter's voice. Its raw, hoarse insistence isolates the spirit. Only when the sun's dominance is restored and temperatures edge toward thaw, do we understand how winter can bruise the heart.

The warmth of sunlight on an upturned face is perhaps the purest joy known to northern peoples. No match for such generosity, winter's fortress shrinks under the tidings of a beneficent sun. Yet it's not a smooth continuum. A Sunday in March can be fifty degrees above zero and twelve below on Monday morning. New snow over old melt punctuates the calendar's advance, threatening walkers and runners with late-season falls and the fated diagnosis, hip fracture.

Geese, cranes, and ducks overhead and in the fields, seed catalogs in the mail, and the smell of a neighbor's dog yard herald a cascade of changes. For highway commuters, driving with the window open, an elbow propped in the slipstream, confirms the shifting realities.

Beyond a windshield pitted with gravel strikes, ravens tumble in marriage rites above the spruce. At the feed store, "Time to Order Bees and Chicks" joins "Breakup Boot Bargains" on the reader board. In the distance, a snow-covered field glints like a mirror. The great loosening is under way. Gravity calls and water answers—burbling, dripping, percolating—running to the sea. Humidity climbs, static electricity falls; the cabin door begins to stick.

On south-facing slopes, snowmelt carves channels in rural dirt roads. Heavy with loess, the coffee-colored runoff drops through miniature cataracts, swirling in pools and eddies before flushing out Cripple Creek to the valley floor. At the level of the Tanana River, meltwater erodes the ice, its assault hastened in the lengthening hours of sunlight. In the shadows, ice holds firm.

Pausing on the trail to Happy Siding, I'm startled by the season's urgency. Three black-capped chickadees suddenly swarm me, as if I were a maypole. Around and around, the trio converges in a flurry of tiny wings, all chittering like scolds, driven by reproductive hormones and territorial claims. Their spiral is so tight, the chase so swift and sustained, I think of throwing open my coat and trapping all three.

Snow berms along the highway give up secrets now on a daily basis. A roadkill, likely a fox, is melting out of a temporary crypt. Ravens have found the spoils and convene a raucous gathering over the sacrifice. With one eye on traffic, the other on bloody sinew, ravens work the carcass with animated determination. Its cause of death poor timing or a distracted driver, the fox shared the ravens' competitive impulse, seeking advantage by stealth and speed until the day it benefited the scavengers' hunger. Picked clean in a day or two, the roadside victim is alms for the birds, carried aloft on iridescent wings.

As the sun climbs, ground-heating swells back into the river valleys of interior Alaska. By mid-April, the first cumulus clouds can be seen building near Tanacross, 150 miles east of Fairbanks. The muscular white towers appear in tentative ranks, soon gaining strength from evaporative snowmelt and bright, windless days. Before the last of the snow withers away, gray-bellied thunderheads advance across the uplands. Distant lightning exposes each storm's violence. Afterward, dark with rain, virga sweeps the horizon, merging with glacial silt scoured from islands and riverbanks and carried aloft by the wind.

In old aspen stands above the floodplain, fairy slippers are poised to bloom. Boreal orchids, they open soon after the snow's retreat. The tiny plants require cold, undisturbed soils, classifying them as global warming indicators. Called Calypso, for the nymph who delayed Odysseus's return from Troy, the yellow-throated flower has a scent that transcends myth, seducing all who venture near.

Just beyond the kitchen window a curious clot of snow in a low-hanging alder captures my attention. A closer look reveals a cloud-white ptarmigan, frozen to death in winter's bitter chill. Still clutching its final perch, the creature died a witness to my domestic routines. Swaying in a warm April breeze, the alder shoulders the corpse in a buoyant wake. Though its winter camouflage is out of date, the bird appears on the verge of bursting into flight. Even as a breath of wind stirs feathers below its bowed neck, the creature is already flying toward the sun. Processes in the alder are preparing to accept the bird as a gift. Humus for the soil; blood for the spirits.

Like ravens and roadkill, the alder and ptarmigan help turn the wheel of existence. Distinctions between life and death resolve in a pattern of reciprocity. As boundaries recede and knowledge advances, it's difficult to say anything is dead. John Muir's assertion that everything is connected to something else is proving metaphysically and mathematically precise.

Yet here on Ester Dome such abstractions pale before the smell and feel of the new season. Imminence is in the air. Beset by irresistible forces, the land quivers with anticipation and release. As winter's logic dissolves, spring equinox conjures the potency of change and asks us to join the dance. When you hear the music, swing your partner.

Karen Tschannen

Two-Part Invention for Winter

My love, am I inventing us? Is there a formula
for the reinventing of love? Or a mechanism
that produces some transfer of warmth?
Perhaps you cannot understand—
winter love is necessary love. In summer
we are free to choose. In summer
there is a whole thick world to love us,
a whole explosion of leaf and smell, freed water,
a whole drench of colored sound.

In winter
there are only sudden, too-sharp clarities,
India ink drawings, brittle sounds of deep lake ice
fracturing, and the fine-thinned pleasure
of white birch against snow. In winter,
absence becomes exquisite, honed down
to lie cold against a length of nerve,
the way thigh aligns with thigh, the way
my palms are pinioned for this moment.

Kaylene Johnson

Ghost Bear

I WENT LOOKING FOR BEARS and I knew where to find them. One of several trails radiating from a nature center ten miles from my house had recently closed because of bears feeding on the late run of red salmon in Eagle River. Surprise encounters along the heavily wooded trail in years past had ended unhappily for both people and bears. The route along the river bottom was cordoned off with yellow tape like the scene of a crime. But the upper trails remained open and I headed for the beaver pond. The area was fairly open with a good view of the surrounding mountains. The water below the beaver dam teemed with salmon. Signs posted along the trails warned that a brown bear had been seen feeding in the area earlier in the week.

I went unarmed while back home my house looked like a war zone. Ammunition was stacked on the kitchen counters; camouflage clothing littered the living room floor. Open gun cases gaped like hungry mouths waiting for the variety of weapons that would accompany my two sons, my husband, a brother-in-law, and two friends into the Alaska bush. They spent months planning this annual harvest and would spend the rest of their lives remembering it. Stories of their exploits in the wilderness grew legendary around the campfire and the kitchen table. Like those of warriors home from the battlefield, their stories formed epic tales of grand adventures.

I could not, however, comprehend their quest. Perhaps they possessed some primal desire to wrestle with the wild and come out victorious. They found satisfaction in providing meat for the tribe and from taking sustenance from the land. From a gardener's perspective, I supposed I understood this. But I was an outsider to their delight of conquest, and the lens through which I attempted to understand was veiled by sensibilities as mysterious to them as their hunt was to me.

This year, along with moose and caribou, animals that provided meat for our family year-round, the guys would be hunting for brown bear. Their special subsistence permits allowed them to hunt for the meat, but not the trophy value of the bear. If they killed a brown bear, they would not be allowed to keep the claws or head. They could keep the hide and were required to pack out the meat. Although they would honor the law, I knew the guys didn't care about the meat. They just wanted to match their hunting skills against a predator of intelligence and prowess and mythological strength.

Maybe I went looking for bears out of protest.

Our sons, Erik and Mark, had grown tall and muscular, and were crack shots with a rifle. Their interest in the outdoors now included my husband Todd's love for hunting and fishing. As I watched them prepare for their adventure, joking with each other about who was the better shot, I remembered sitting around a campfire when they were youngsters. Mark had a jack-o'-lantern grin, his mouth full of gaps and adult teeth too big for his little-boy face. Erik, serious and contemplative, sat leaning against me with easy affection as we roasted hot dogs over the flames. That moment, with the scent of wood smoke in the air and with the whispered song of a creek in the background, I felt utterly, unalterably connected to these boys. They were my homeland, the port at which all good and meaningful things were anchored.

Their joy for this fall's hunt was now so outside my grasp, so beyond my sharing, that they had begun to seem like a foreign land to me. The boys I thought I knew so well were disappearing into a territory that seemed to me shadowy and uncertain.

Mark, seventeen, sported a wild mop of brown hair and intense, brooding eyes. He claimed that hunting moose and caribou no longer posed much of a challenge. Given the opportunity, animals of prey run away. But a bear. Now that was different. A bear was a predator and could kill a man with the swipe of a paw. Mark would have preferred to use a bow and arrow rather than the 30.06 that he was assigned for the hunt. Part of me admired him for wanting to even the odds—as long as someone backed him up with a big bore rifle. A larger part of me wanted Mark to put the animal in his sights and then lower the gun and let the bear eat berries in peace.

Erik, nineteen, was more philosophical and ambivalent about the prospect of killing a bear. His blue eyes narrowed as he thought about the upcoming hunt. "A bear is at the apex of Alaska's ecosystem," he said. "It seems to me that on the evolutionary scale, all other species of animals and plants are lesser beings. If there is a moral dilemma about hunting for bear, it would be the act of taking the best thing that nature has to offer."

I wondered what he would do when he put his sights on brown fur shimmering in the evening sun. Of course he would have his brother, his dad, and the other hunters to answer to—guys whose second favorite pastime after hunting was the sport of ridicule.

Todd sees bears as direct competitors for the moose and caribou that he himself wants to harvest. He hunts bears out of a conviction that wildlife mismanagement and the lack of predator control have created an ecological imbalance, diminishing the supply of game meat. He will shoot a bear to save moose and caribou from losing their calves to an overabundance of predators. Bears account for about seventy percent of moose calf mortality. If moose and caribou have the opportunity to grow up, they will produce more calves and thus more game to hunt.

My feelings about bears are a mix of awe and wary respect. Bears are beautiful, graceful animals whose fierceness in defense of their cubs I admire. Their presence in the mountains and backcountry and in Eagle River Valley where I live reminds me that humans are not the all-powerful creatures that we like to imagine. I am not naive. I know that bears pose a potentially lethal risk to people—including my own family and friends.

And yet, the endless hours of preparation for the hunt—the boxes of shotgun shells and rifle bullets, the mounds of tents and sleeping bags, the ropes and tarps and other accoutrements—all seem excessive to me.

It annoys me the way taxidermists pose mounted bears. Invariably the bear stands on hind legs, with lips curled to reveal fearsome teeth, and with claws extended. As a matter of fact, bears stand on their hind legs when they are curious and want a better look at something. The stance is not one of aggression. The bear was most likely shot as it grazed along a sunny hillside, not as it tried to attack the hunter. But suppose the bear did rise up on hind legs before charging. Unless it is hunting season and you are licensed to kill a bear, state law requires that anyone shooting a bear in defense of life or property must skin the animal and relinquish the hide, head, and claws to fish and game authorities. The hides are then tanned and sold at auction. And supposing that someone could legally keep the cape of an attack bear, I find it hard to imagine that anyone would want the reminder of a terrifying bear encounter posed in the corner of the family den. Most snarling mounts of bears are fabrications, poses that imply that the bear was aggressive and that the hunter was brave.

Bears, universal characters in many Native American stories, represent qualities of power, healing, and gentle strength. The Tlingit people say that the bear Kasha is the Great Mother who gave birth to all animals. Because of their humanlike qualities, bears were believed by many West Coast Natives to be Elder Kinsmen. When killed, the hide of the bear was taken to the chief and treated as a high-ranking guest.

A Native Gwich'in woman once told me about a family of bears she and her brother encountered one spring when she was in her early twenties. Velma Wallis, author of *Two Old Women*, *Bird Girl*, and *Raising Ourselves*, is a short, cherub-shaped woman with high, arching cheekbones and a puckish nose. Her eyes crinkle into curved slivered moons when she smiles. But as she told this story, she was not smiling, not even twenty years later.

Her older brother had been drinking, and as their flat-bottomed boat roared around a corner of the river near their home on the Yukon River, they startled a mother black bear and her two cubs. The cubs scampered up a tree and the mother disappeared into the brush. Her brother idled the engine and grabbed his rifle.

"Don't shoot the cubs," Velma protested. He ignored her, took aim, and squeezed off a shot. Velma screamed as one of the cubs crumpled and fell from the tree. Her brother laughed at her.

"What I didn't know at the time," Velma said, her voice growing quiet, "is that I was pregnant."

Tears welled in her eyes. "Not long after this happened, I miscarried."

She paused and looked at the napkin she had folded, and unfolded it again. "I know it sounds crazy, but I'm convinced that mother bear took away my baby, just like my brother took away hers."

The first time I went looking for the bear, I brought our dog, Sue, and watched from the trail. I looked for the closest tree I might climb and calculated how long it would take to climb it. I mentally measured the distance between the dam and the trail where I was standing—maybe fifty yards. Staying on the trail, I stood behind tall stalks of cow parsnip so that I was not so obviously in view should a bear arrive on the scene.

Sue grew confused at my stillness. She was used to rigorous, extended hikes. So far we'd only traveled a short mile. She trotted to the water, drank, and came back to stand beside me. With mild interest, she watched as two beavers played in the pond, pushing folds of glassy water as they swam. Although she was an agreeable companion, with an uncanny sense to leave wildlife alone on our excursions, she distracted me, drawing me into a familiar state of maternal concern. I could climb a tree, but she couldn't. Besides, no matter how benign her presence, she might unwittingly agitate a bear. After half an hour without seeing anything but the beavers, we headed home.

When I returned to look for bears without the dog the next day, I wandered off the trail and sat on a log stump. A park ranger stood on the trail where I had earlier watched. He carried a shotgun and spoke intermittently on his handheld radio. Hikers occasionally stopped to visit with him. When he wasn't talking on the radio or to passersby, he seemed restless, perhaps nervous. The sound of his boots scuffed on gravel. He checked the squelch on his radio. Checked his gun. It occurred to me that he didn't really want to see a bear.

As I sat and waited, I thought about the guys and their upcoming trip. I had no qualms about hunting as long as the animal was an abundant resource in the ecosystem and was treated respectfully both before and after its death. I could understand my husband's and sons' call to the outdoors during the riotous autumn of the year. And I appreciated the value of wild organic meat. Yet I knew there was something more to the killing, something primal and raw and incomprehensible that drew my husband and sons to the hunt. While the gun I once carried had been a means of protecting my children, it became something altogether different in their hands. Something entirely mysterious to me.

Eventually I got up from my log seat and went to talk to the ranger. I was curious to know when the last sighting had been and whether there had been trouble.

"No trouble," he said. "Not yet, anyway. Backpackers have been seeing him every few days and we're worried he's getting too used to people."

I asked him about the shotgun and what he intended to do if he saw the bear. He explained that he had several types of rounds for the shotgun. The goal was to teach the bear to avoid humans. The first shot would be a "cracker" round, a shell that transmitted a flash of light with a loud, crackling noise. The second choice was rubber bullets, a painful but not lethal reminder that people meant big trouble. Finally, the ranger had shells loaded with slugs, just in case a confrontation turned ugly and he needed to kill the bear.

I asked if he had ever killed a bear, and he said no—but that he'd had some close calls he didn't much care to think about. That evening the bear stayed away.

Alaska is home to three species of bears. Polar bears reside north of the Arctic Circle. Black bears come in a variety of colors ranging from light gray (glacier bear) to cinnamon brown and black. Brown and grizzly bears are names for a single species of bear. Home to ninety-eight percent of the United States' population of brown bears, Alaska is one of the few places in the world where brown bears can be hunted. The concentration of browns in central Alaska is one bear per fifteen-to-twenty-three square miles. On Kodiak Island the concentration is nearly one brown bear per square mile. The island's abundance of salmon also gives Kodiak bears the opportunity to gorge and grow inordinately large. It is here that many hunters seek out big game trophies.

Jim, the husband of a friend, told me about his bear experience on Kodiak Island. He had gone hunting with friends, and on the last day of the hunt he spotted a brown bear on the hillside several ridges away. It was getting late in the day, almost too late to shoot a bear. It would take hours to skin and pack out. Besides, he and his friends needed to reach their boat before dark and it was anchored several miles away in a secluded cove.

Jim remembered looking at the bear again and deciding it was just too good an opportunity to pass up. The hunt was nearly over and they still didn't have their trophy. This might be his only chance.

It took Jim longer to stalk the bear than he anticipated. His first shot wounded the animal and it stumbled into a stand of alders below the tree line. Jim hesitated. It would be foolish to follow a wounded bear into the brush. He also knew that, after hearing the rifle shot, the other hunters would be on their way. So he mentally marked the place where the bear had disappeared and went to meet them.

By this time Jim had hiked many miles over steep terrain and hadn't eaten since breakfast. The exertion and adrenaline of the hunt had depleted his last reserves of energy. Shortly after he met up with the other hunters, and as they returned to find the bear, Jim suddenly collapsed.

"I fell flat on my face. It was like a marathon runner that hits the wall," Jim explained. The hunters stopped and let Jim rest and drink. They offered the last candy bar from someone's pocket. As the evening shadows began to lengthen, they resumed their hike and found the bear. It was dead, lying not far from where Jim had shot it.

The trek back to the water's edge became a struggle as they slogged through a knee-deep stream under the weight of the bear's hide. Adding to the tremendous exertion of the day, the chill of evening air and icy water seeped into Jim's bones. Every step became an act of sheer will; his movement seemed mechanical and forced as his limbs grew heavy with fatigue. He found himself beginning to stumble, a sign that he was growing hypothermic.

When they reached the cove, Jim suggested they leave him behind. The other hunters could move more quickly without him. Once they got to the boat, they could motor over and pick him up. He would build a fire as a marker along the dark shoreline.

The others agreed. It wasn't that much farther; the boat would be just a few miles down the beach. Within a couple of hours they'd be back. They left and Jim began collecting wood for the fire. It had rained earlier in the day, and the only dry wood was what he could find by crawling under the branches of broad spruce trees. He finally had enough kindling and dry grass to start a fire, and it was a good thing. He needed to get warm from the outside; his own body wasn't generating much heat.

With cold, trembling fingers he reached for the lighter. But his pocket was empty. Quickly he tried the other pockets. The lighter was gone. Either it had fallen out as he bent over to skin the bear or it had dropped out as he scrounged for wood. He tried to retrace his hunt for wood, but found himself shivering and growing disoriented in the dark. Rain began to drizzle.

He realized with sudden certainty that he was now in a survival situation. He had no fire or food. All he had was a blood-soaked bear hide.

A bear hide.

The very bear whose life he had claimed, the bear that had culminated a great hunt, would now be the bear that would save his life. He unfolded the hide and rolled himself in it. Whatever part of Jim's clothing was not already wet now became saturated in bear blood. But the hide was warm, warm enough to keep his teeth from chattering. Fitfully Jim slept in his womblike bed, dreaming bear dreams as he listened for the sound of a boat motor. He would not hear it until dawn of the next morning.

The third time I looked for the bear I went later in the evening. I positioned myself on the other side of the pond about one hundred yards from where the salmon, weak from their upstream migration to spawn in these waters, splashed below the crosshatch of wood that formed the dam. I sat behind a fan of exposed roots from a fallen tree. This time it didn't occur to me to look for an escape route.

I cannot explain my lack of fear. With each visit to this place, every nerve ending seemed to grow more aware of my surroundings. Even my uneasy thoughts about the upcoming hunt faded with the gentle echoes of evening light. I noticed the veins and capillaries of individual leaves on crimson cranberry bushes. The quiet stillness amplified the trickle of water over the dam. The smell of the air changed from the scent of grass to the tang of ripening berries and then to the cool musk of dirt and shadows. For over an hour I sat and watched. Eventually, even the whisper of bird wings grew quiet, and cool mist began to rise off the pond.

Then, like a ghost, he appeared. On the other the side of the dam, the brown bear ambled into the middle of the stream and sat down in the water facing me. I don't think he saw me, but if he did he was unconcerned. He dunked his head and then looked up. Water trickled off his ears and snout. He was sunny brown, an adolescent bear, maybe two years old, around four hundred pounds. Still sitting, he lifted his front paws out of the water and then, comically, he allowed himself to fall over backwards. He rolled over, got back on all fours and then sat down again and watched the salmon. He didn't appear hungry or intent on anything in particular. He was lean and lanky, gorgeous and just a little bit goofy looking. Then as quickly as he appeared, he wandered off, his motions fluid and effortless and quiet.

Had I not been anchored by the weight of my own body, I might have flown with euphoria, lifted up and out of my skin into the cool evening sky by consuming joy. How wonderful to see this bear, undisturbed, playing almost humanlike with relaxed abandon. I stayed another hour, hoping he might return. Twice more I saw him cross the stream, each time a little farther away. Seeing the bear again proved he was real and not some wisp of wishful thinking. Then somehow I knew the bear was gone for good and it was time to go home.

At the dark end of dusk, the crunch of my footsteps on the trail made real my own presence here—a place as tangible and mysterious as the spaces of family and motherhood and the conflicted human world in which I lived. Yet for this moment, the taste of crisp night air eclipsed all else except the rising moon to the north.

Alexandra Ellen Appel

Light on the Kuskokwim

Breakup #3

I. What is it that gives you away
is it a strand of your long dark hair
a shadow in your voice
coupled with my longing
I see you
all those ancient times
at sea in a skin boat

I sew and sing the rhythm of our drum
these are the secrets
we will not share
even though our bones ache

I tasted the water
once
is this enough, walrus man
wolf claw man
mask maker man
spirit keeper is what they called you,
once.

II. I've called you.
From the tip of the ice ledge
watched the sea
try to claim
what belongs to Spirit
you exist in memory
in dream
translucent, the light returns

the taste of salt
water in my mouth
is dream, is salmon
and seal and caribou
is myth.
I stand at the mouth of the river and call you home.

We make fire from driftwood and dry kelp.

Steve Kahn

Standing on a Heart

THERE ARE PLACES ON THE TRAIL that hold ghosts. Places that haunt and beckon us. Trees or fields, ridges or rocks that trap us. Forever dust and deadfall, the slow snake of path ahead, curving, shrinking into nothingness. For me, one such place lies where an old jeep trail that links Farewell Lake to Farewell Station passes near the west end of John Lake. As small patches of boreal forest go, it is unremarkable. If you are walking with the morning sun at your back, something eases you down the rutted track. A gentle downhill, and an earth pull that makes you believe you could walk on endlessly. Dwarf birch gives way to tall grass as you approach the stand of white spruce hunched at the bottom. To the right the conifers are larger than most in the area. A deep game trail winds under their boughs, up from the edge of the lake. You can't see the water, nor how the lake wraps itself, moat-like, around three sides of a steep ridge. But you feel it. I have passed this place in the boot-sole sink of spring, known the dust and mosquito buzz of midsummer, the crisp flirtation of autumn, and the cold that follows.

In my history here, there is a sadness too. Even the red squirrel's bark loses its edge as the sound weaves down through the boughs, more a lament than a scold. Spruce cone detritus forms random, soggy piles beneath the tallest trees. There is a graying at the edges of things: the gnarled roots of an alder, a withered bolete abandoned on a branch, the shape of my hand.

It was the late 1970s and I was doing what young, single, wilderness lodge caretakers did in Alaska during the winter: trap. There was a rhythm to most days that I came to expect and love: each morning's first sensation of cool air on my face; looking up to the swirl of metal springs on the underbelly of the upper bunk; a glance to the cassette deck mounted on a wooden shelf, surrounded by the little boxes containing my favorite tunes, all within arm's reach; my left shoulder leaning against the chestnut-hued log wall. And, as if I were the one to introduce sound and smell to the world each morning, the creaking of the plank floor as I walked over to the barrel stove, the crackle of dried spruce, followed by the perk and aroma of coffee. In many ways the setting was idyllic. It seemed I was finally leading the life that I had dreamed about for so many years.

What I knew about trapping came from reading and from casual comments from outdoorsmen I'd met. My father, an avid hunter and fisherman, didn't want any part of trapping. He'd grown up on a farm in Wisconsin, and through most of his teenage years his parents raised mink. When his brothers left home to serve in the military during World War II, he was the one who fed the mink and cleaned their cages. He broke their necks and skinned them. He smelled the strong odor of confinement and felt the pain of their teeth sinking through leather gloves into the meat of fingers.

Under the hiss and brightness of a gas lantern I studied my dog-eared trapping manual. In the field, I studied harder. Much of what I read about animal behavior turned out to be true.

Beneath the protecting branches of a spruce tree, I spiked a stout, dry, seven-foot pole that ran diagonally from the ground and protruded eighteen inches past the tree. Called a pole set, it is a favorite of those working marten country. The trap and bait are placed on the upper end, and the marten, a tree climber, ascends the pole to reach the bait and consequently steps into the trap. The trap is secured lightly to the pole with a string, wire, or bent nail, and in the animal's struggles it drops off and swings clear, taking the marten with it. In cold weather the animals freeze quickly, usually in a few hours. That is how I had always found them. Dead and frozen.

But a Chinook wind had warmed the December air above freezing. There was no need to stoke the barrel stove that morning to ensure a warm cabin upon my return. The strap-hinged, cotton batten–chinked door latched easily. Outside, my snowshoes rested on the brow tines of a huge set of moose antlers spiked to the front wall. I would not need the webbed wooden frames on the hard-packed trail. I shouldered my pack and began walking.

Within the first thirty minutes I passed two sets with nothing in them, but ahead there was movement. Beneath the pole, a marten hung inches from the ground. There was a minklike grace to its form. Stretched out, the animal looked sleeker than the ones I had seen peering down from the branches of black spruce, their soft, round ears framing teddy bear faces. The jaws of the Victor No. 1 single-spring held one front paw, its fine, ivory white claws emerging from the plush brown fur. I unbuckled the belt of my pack, shoulder-dipped and arm-tucked my way out from under the straps.

This was supposed to be my moment, a chance to confirm my membership in the fraternal order of mountain men. An opportunity to collect a token of manhood and some tangible proof that my time had not been wasted. Justification of existence on a primal level. But there was no thrill or sense of accomplishment, just a nagging pressure at my temples and a tightness in my chest.

The marten's efforts to escape came in waves: it was an amateur gymnast on the rings struggling to perfect a new move, then swinging to rest. I had read that a sharp blow with a stick across the nose would stun a small animal and it would die quickly once you stood on its chest. The kicking and squirming of the marten intensified when I approached it with my walking stick half-cocked. The constant motion, the twirling of the furry mobile into new positions, made me check the first swing and the second. Finally, I choked up on my surrogate club for more speed and accuracy and rapped the marten in the face.

Whether the blow was off-angle or merely not hard enough, I don't know. The trap chain that was wired to the pole slipped several inches, and the animal's hind feet touched the ground. It lunged at me, jerking against the steel jaws that held its paw. I jumped back. My only recourse was to flail away with my stick like a man snuffing out a burning bush, until the marten lay there, stunned and twitching. I stepped forward quickly, one foot on its chest.

There was a taste of iron in my mouth. *Please die, please just let me kill you.* I cursed the steep and messy learning curve. I reprimanded myself, thinking I should have secured the trap chain better, positioned it farther up the pole so there was less chance of the marten touching the ground. I thought of the far too many squirrels and gray jays that had perished before I learned ways to hide my traps. And still, even with increased experience, there were unwanted deaths, remnants of fur, feathers, and blood.

We can learn a lot from reading, but there is nothing like experience. And there is no chronometer sensitive enough to measure the length of certain moments. Like the time I knelt a dozen feet away from a trapped lynx, wiring a snare in the shape of a noose to the end of a long pole, and the only sound was the squeak of dry snow against my boots as I stood. Because a bullet hole will bring less money for the skin, because lynx die easily. Cat eyes upon me, the loop of wire slipped over the feline neck. Lift up, up, I told myself, and the full weight of its death was in my hands.

What I didn't learn from reading trapping manuals is that finding my place in the life and death cycles on which all natural systems depend would be hard. Courage to change comes in increments, in admitting the difference between killing for food and killing for fur. For a while I tried to bridge that gap by eating the flesh of muskrat, beaver, and lynx. But I was eating the meat because of a self-imposed obligation to minimize waste, not because I relished the flavor. Though my culinary limitations were a factor, I never got past the greasy, sweet taste. Even though I sold the fur, I couldn't shake the thought that these creatures should be more than a commodity.

I quit trapping. Years passed and I came to live on another lake in remote Alaska, surrounded by different mountains. A neighbor to the east fed and photographed the animals; another to the west trapped and skinned them. One winter when I watched the easy gait of a westbound wolf, my heart lifted, then too quickly settled. I never heard the shot, but the voice of my neighbor was clear as he boasted over the VHF radio how he never had to leave his cabin, just picked up his rifle and opened the window.

Later I passed by the crumpled carcass, misshapen, and bird-picked on the ice. The cold I felt was bone and muscle deep, and the ice I stood on was not the only thing that had buckled and fractured.

I continue to hunt and fish for food, but something has changed. Salmon and burbot are filleted closer to the bone. Grouse blood stains fingers that hold the bird a bit longer to feel its warmth, its softness. Its feathers, its feet, its weight are all a part of me, and I think of my parents, my wife, and my friends. As if joy were to the east and sadness to the west, I stand on a hillside exactly in between, feeling both.

The experts tell how to trap a marten. What they don't mention is if the snow is too soft or deep, the body of the marten beneath the foot settles away from the crushing pressure. That it is necessary to put even more weight on that leg to finish the job. They didn't tell me whether the beating was coming from under my foot or from my own chest. Or how someday I would sit next to my mother in the hospital waiting room while my father's guts were in the hands of a surgeon. How my mother's heart, for decades enlarged and beating irregularly, would feel as if in my own hands. I look out the window at the mountains and imagine a small creature held by steel jaws. I remember a tiny patch of earth just west of John Lake and how nobody told me, as I stood on that heart and stared straight ahead, that the pulse of the moment would last a such long time.

Jerah Chadwick

Cormorant Killer

I want to believe the old stories
that you will come back
as what you wasted: aurora
of black swan neck, seaworthy wings
bearing you to/from your chicks
in the rock, along the cliff-

faced yawn of some beach
where a middle-aged man waits
out his free time, whiskey
in one hand, already drunk
with the gun in his other.

Shoot one and its mate
will soon come searching
he brags at work of the weekend
trips, blasting the same
empty bottle he keeps becoming

as shards of your feathers fall
and you fly again, snake raven
for fish or with gullet full.

May it always be nesting season.

R. Glendon Brunk

Facing East

AT THE VERY SOUTHERN EDGE of the development my wooden observation tower sat alone out on the tundra. I entered it via a wooden ladder propped beneath, popping up inside like a gopher through a trapdoor set in the middle of the floor. The view south through the Plexiglas windows was of tundra spreading endlessly, a dull green, polygonally patterned plain pocked with countless small black lakes. On the perimeters of the lakes migratory waterfowl still nested—black brant, tundra swans, and pintail ducks. On the drier ground, amid stunted willows, golden and black-bellied plovers and Lapland longspurs hid nests from the relentless predation of long-tailed and parasitic jaegers. Occasionally an Arctic fox wandered by, or a snowy owl floated past like a winged ghost. On still days I could hear the plaintive cry of a yellow-billed loon above the sounds on the road. Far out across the tundra it called, signaling like a prophet.

My view of the other three directions was significantly different. "The slope," as it's called by workers there, was even then, in 1989, the largest contiguous industrial development on earth, several hundred square miles. (It's close to double that size today.) The oil companies boldly touted just how carefully they were developing the Arctic. But the reality was and is that petroleum production is a toxic, destructive, go-for-it business, dedicated to one thing only: profit, billions and billions of dollars taken at the expense of the earth.

The slope told the tale of it. Toxic waste bubbled in settling ponds and seeped into nearby lakes. Huge holes, where billions of yards of gravel had been mined, gaped like open sores on the landscape. Every turn of the horizon was broken by the works of man: a maze of pipelines and roads, drilling platforms, radio towers, transmission lines, camp buildings, oil wells, refineries, and production facilities. There was something surreal and sinister about it, a space odyssey quality that confused the mind, a hard, angled, technological sterility. Being an observer of wildlife there, of observing living things juxtaposed against everything that was their antithesis, was a schizoid experience.

My part in the caribou studies was not glamorous, just interminable hours and days of waiting and of scanning an often empty tundra plain with my binoculars. There were only my idle thoughts to keep me company. Behind it all was the ceaseless blowing of the north wind, careening off the tundra, rattling and shaking my tower so hard at times that I thought it might tear the guy wires from the frozen earth and I would sail away over the white expanse of the ocean, like some reluctant space traveler. In time the north wind became part of my subconscious. But then one day it would suddenly die and the world would become loud with its absence. That was the moment when gray arctic spring yielded to the heat of the round-the-clock sun, and summer began. The time when the tundra world instantly became an angry, buzzing mop of mosquitoes and flies.

Like some daunting apparition the caribou would suddenly appear, a line of brown velvet antlers moving far off through the heat waves. Then they would take form, maybe ten or a thousand, moving quickly and determined, driven to the edge of panic by the bugs. They were headed for the cooler, mosquito-free shores of the Arctic Ocean that lay just a few miles to the north. This response to the torment of insects is a migration pattern that has gone on for as long as there have been caribou in the Arctic. But times had changed.

My job, if they came within my study area, was to count their numbers, map their movements, and note their reactions to the elevated pipeline and to the road adjacent to it that ran along behind my tower. "Success rates crossing pipelines" was the scientific jargon we used, as if caribou were competing for something.

If I was on the ball, not lost in some far-off fantasy, when they appeared I would snap to attention and grab my binoculars and the four-finger hand counter, and begin clicking numbers. The caribou always came too fast. Stay calm, I'd tell myself. The sure way to blow a count is to panic. Also keep in mind vehicles on the road. Vehicles always made the difference.

I would watch them cross the outer fringes of my study area, note the time, and check for vehicles on the road. Too often the same maddening scenario would develop. The caribou closed on the pipeline. A couple of hundred feet away the lead animal, usually a Roman-nosed old cow, would throw up her head and stop. The others would immediately halt behind her, begin milling and shaking their heads, twitching hides and tails, battling the mosquitoes. Then very cautiously the lead cow would advance toward the pipe, her nose held high in the air, like a bird dog on fresh scent. She would stop again, clearly mystified by this odd form looming over the tundra. Then a pickup truck or a belly dump or some other piece of heavy equipment would approach noisily on the road. The old cow would turn and begin paralleling the pipe, the others following her closely. The vehicle would blast by in a cloud of dust, more often than not with its horn blaring. Then it would happen: the herd would turn as one and run, eyes wide, nostrils flaring, stampeding back toward the horizon from which they had first appeared minutes earlier.

What I remember most is the anger, the frustration I felt when I watched the vehicles approach and the caribou fail at crossing. Each day the anger built and seemed to have no place to go. So many times I stood tense and hopeful, then watched the same scenario unfold, felt their defeat as the animals turned away. And with it I felt and began to actively articulate to myself a much larger defeat, the lunacy of this clumsy, arrogant game we moderns play with the natural world.

Over the course of the three summer seasons that I participated in the studies, my colleagues and I noted some things. Gravel ramps built over lowered sections of the pipelines as crossing devices failed miserably. When confronted with any obstacle foreign to many thousand years of arctic experience, caribou in numbers usually became confused and retreated. Given enough time, though, some would eventually figure it out and begin to cross under pipelines if they were elevated enough off the ground. The bulls would duck their heads in an exaggerated way, careful of their antlers, and then quickly shuttle under. But trucks and machinery on the roads—"vehicle interactions," we called them—would almost always foil any attempt. Bulls and dry cows clearly got through the oil fields more easily than cows with calves, which rarely succeeded, even with no vehicles on the roads. A few bulls did adapt to the oil fields, in fact seemed to prefer them—most likely, we speculated, because of the lack of predators.

My colleagues and I did not record the confusion, the white-eyed panic, the separation of cows from calves when pipelines and roads with traffic were encountered. We did not record the casual attitude of the workers in the oil fields, their lack of interest, or even their outright disdain for living things. We did not record the caribous' beauty, their comic grace, the miracles of minute ecological adaptations that allow them to thrive in a punishing environment. Nor did we record the anger that some of us felt. All that we recorded was what was allowed within the narrow parameters of the study. We generated "data" to be run through the mathematical hieroglyphics of computer models, data that were then spit out as statistics, graphs, charts, and technological language devoid of any smells, colors, or feelings, all to be thickly bound in reports to gather dust on desks in Fairbanks, Anchorage, and Houston. Somewhere early on in the process, at the tip of my own pencil, caribou quit being living things, with all the miraculous interactions of any species, and they became an abstraction.

I felt frustrated by the parameters. I wanted other people to feel what I felt, to observe what I observed. I wanted them to know what it was like to see the Arctic change so rapidly, to have seen this same tundra world before development. To have seen it wild and unscarred, and then a few short years later to see the heart torn from it. When I expressed my concern to one of my colleagues, he reminded me that scientists pride themselves on being detached, on being objective. "You can't care about these animals and do the work," he said. He was the same one who called caribou "tundra maggots," referring, I guess, to their numbers and the way a large herd on the move seemed to squirm in the distance.

I could not help but witness that many research biologists in the oil fields go numb. The reductionist nature of the "science" they practice, the crush of bureaucratic and corporate demands, of seeing over and over again the destruction of wild places and of living things—all these things divorced some from their feelings. I know, though, beneath it, at one time most cared. Almost all the field biologists I have ever known are people who began their careers with a keen appreciation of the natural world. The consulting company I worked for was as reputable as they get, committed to good science, at least the kind of science that was required by governmental agencies catering to corporations tenaciously dedicated to getting what they wanted. I know, too, that most of us working there were uncomfortable at some level with what we were doing. But our reasoning went, if we don't do this study, some other, less ethical consulting company will. At least we'll do it the best that it can be done. And beneath that reasoning was one ever-insistent practicality: we all need to make a living.

Yes, so many things we do in life are justified by the dual rationalizations of inevitability and economics. The voices of modern reason. By my third summer working in the oil fields, the voice of reason that I had employed was sounding more and more hollow. I couldn't help but feel that most of us working there were giving away something important when we sold ourselves to the oil corporations. It's hard to say this now, hard to make judgments about people I still know and care about, but there was a heavy price to it. Something died in all of us, some deep part, some vital idealism and passion that feeds the soul and gives us cause for full living.

When hearts are unengaged, any work becomes a paper cutout. Wildlife biology (or any science, for that matter) becomes an exercise dominated by technicians, computer-model addicts, people who have bought the notion that scientific practice is supposed to be completely value-free. Which of course it can never be. Because there are humans doing it, with all their human biases and value-based perceptions.

It bothered me that the old-time naturalists/biologists, those men or women who dedicated their whole lives to understanding intimately a place and the interrelationships of living things in that place, the biologist that I imagined myself being when I entered college, were no longer accepted. The patience and deep caring of that kind of observation was missing in everything that we did. Everything was speeded up, computerized, depersonalized, abstracted. It's become even more so today. Too much of biology has become the questionable science of risk assessment, an exercise dedicated to answer only this question: How far can we push a species with our industrial activities before it will fail entirely?

On the North Slope we took the science of abstraction and risk assessment to a new level. We studied caribou knowing full well that nothing would change in the oil fields if our study suggested harm to caribou populations. Our only intent with the study was to document what was already under way, to blanket with statistical jargon what anyone with a lick of common sense could see was a growing disaster for wildlife and for what was once a wild place. It was already well documented that cows had ceased calving in a major portion of their traditional calving areas because of human activities. This displacement was concentrating the cows into smaller areas, thus depleting nutrition sorely needed for calving. Speculation was that calves were being born weaker, making them more susceptible to mosquitoes and other stresses. It didn't take a science degree to figure out that cutting off access to mosquito-relief habitat added yet another complication to an already highly complicated existence. The effects could not easily be proved with a tidy, short-term study. All that was needed to see it, though, was a dollop of common sense. All it took was a willingness to look honestly, an openness to feeling.

One of the oil companies operating on the North Slope used to run an ad in national magazines with a glossy color photo of some massively antlered bull caribou grazing contentedly by an oil rig. The text went something like this: "Working to protect our environment while providing for America's future." This catchy slogan used to turn me livid. I knew the truth. I wondered why scientists didn't speak up, go public, tell people what's really going on up in the Arctic.

As my third season drew to a close, I knew that I was in the wrong place, working in the wrong business. Each day I sat in my tower and fantasized about walking east. I saw myself walking quickly, leaving the oil fields behind. I walked effortlessly through the jumble of rivers and lakes, fifty miles, until I came to the broad, braided delta of the Canning River. Beyond the Canning was country devoid of oil structures, untouched and primal. Beyond the Canning lay the country I'd visited years earlier, the Arctic National Wildlife Refuge.

Autumn comes to the Arctic in a rush. One day the tundra is green and the next it's a palette of colors: reds and rusts and muted yellows. Newly feathered waterfowl begin to test their wings, restless, pushed by an inner whispering that warns of the hard hand of winter that will soon grip this place.

In my tower I could hear the same whispering.

It was well past midnight when I climbed down the ladder of my tower. The sun was still alive, a tepid orange orb set low on the northern horizon, casting a diffused yellow light over the tundra, a gentler touch than the pale, angled glare of daytime. I got in the company pickup. I could not go back to my room at the construction camp where I lived. I drove over gravel roads, through a maze of silver pipelines, the capillaries, veins, arteries of the giant. I passed dozens of well pads with rows of metal-sheathed pump houses standing like space-age knights in review; refineries with gas flares pulsing red into the night sky; and portable drilling rigs set on newly laid gravel pads far out on the tundra, towering ten stories high, aliens in a world where the tallest plant might rise a foot. I passed equipment yards with rows of dozers and graders and belly dumps; construction camps with lines of portable housing units connected in close rectangular patterns; and the ordered sprawl of a main operation camp, this one famed for an Olympic-sized swimming pool, a full gym, and a tropical garden inside.

I drove northeast until I came to a long causeway that jutted like a finger into the Arctic Ocean, to a man-made gravel island that held another production facility. I pulled to the side of the road and parked. For most of the year the Arctic Ocean is icebound, covered beyond imagination with a solid white armor. In summer, though, the winds shift from the northwest to the southeast. By August the sun has weakened the ice enough that the winds can push it offshore. It lies far out to the north then, visually just a shimmering mirage, a refracted white band of light against the darker curve of the sky.

Escape. I shouldered a small pack and began walking quickly east. I waded shallow braids of the Sagavanirktok River, then gained the narrow beach along the deceptively placid Arctic Ocean. The beach, a slate gray interface between the tundra and the ocean, was littered with chunks of Styrofoam and soda cans, survey stakes topped with tags of fluorescent red ribbon, frayed lengths of rope, and the ever-present symbol of modern man in the north: rusted, fifty-five-gallon drums.

I walked, zigzagging from the beach out onto the tundra, skirting ponds and small lakes, then back to the beach again. Along the way I spotted an arctic fox busily investigating a colony of arctic ground squirrels, a snowy owl perched motionless on its hunting mound, and out over the ocean the white-on-black flash of king eiders in flight. I must have walked an hour or so before I came to the banks of a black tundra creek, too wide and deep at its mouth to cross.

At least I was out of the oil fields. I turned south and stared out over the tundra, the coastal plain that some call barren. In the distance, eighty miles at least, the peaks of the Brooks Range caught the early morning light. Somewhere in those mountains the new Dalton Highway crossed a high pass, connecting the oil fields with the rest of the world. Its construction had speeded up time on the coastal plain, had brought it square and immediately into the insatiable hunger of the twentieth century.

I stood still and let the sounds of the tundra take over. Close by I could hear the whistled *chu-leet* of a golden plover, and somewhere in the distance the cry of an arctic loon. I felt the wind gather itself, watched it touch the tundra, bend strands of cotton grass, and riffle the surface of ponds.

I faced to the east again. I thought of my fantasy of walking to the Arctic Refuge. I knew that the oil companies, speculating there were paying quantities of petroleum in the refuge, were spending millions on a public relations effort to get permission from the federal government to develop there. It struck me then that I stood at a place on the planet that symbolically focused so many of our modern dilemmas. I stood at the north edge of the world, one of the last places that could be exploited in a big way. To the south, far beyond the Brooks Range, millions of North Americans were demanding more of everything, more goods, more oil, and, ironically, more wild places to escape to. West of where I stood, the oil fields were witness to what we've been up to for over three centuries, the ill-conceived notion of frontier, of resources unlimited, the unquestioned righteousness of industrial technology and an ever-expanding economy.

And to the east, in stark contrast, survived a remnant of unaltered nature, the priceless remains of a former world, the path back to something our overly denatured souls most certainly yearn for. It came to me then that I stood at the interface of choice. Do we modern humans continue to deface the world in order to keep the illusion of our progress alive? Or can we begin to face another direction, begin to make the hard choices of living here in a radically new way?

Awakening. The journey of the heart is an intricate weaving of small moments that build and build to finally turn us in a new direction. As I stood there it also came to me that I had to ask myself the same questions. I had to see my part in the scarring of Alaska, not always an active part particularly, more one of compliance by silence and avoidance. It angered me when I thought of oil development in the Arctic Refuge. But what was I doing about it? How was my participation in the oil fields, my assumptions and consumptions, any different or more righteous from those of the executive who signed my paycheck? If I wanted the guarantee of wild in my own life, and wild places for generations to come, what was I doing about it? If I cared about my own child, what was I doing to leave her a different legacy?

The sun was climbing the east flank of its orbit, casting a new, harsher quality of light across the tundra. I made some decisions then. It would be my last season in the oil fields. I would never again knowingly give my energies directly to any institution that was dedicated to profit at the expense of the earth. Of course, I understood the key word was "directly." I could not avoid the many compromises that each of us lives with each day, the ways that we are forced to participate in the destruction. But I could work for fundamental change. In that moment I decided that I had no choice. I could not run again. It was time to follow the other voice. Whatever it asked, wherever it took me, that's what I had to do.

That moment up on the North Slope turned me in a new direction. Yet in another way it simply verified the course that I'd been on all my life. Like it or not, I was about to come out of the closet, to take a stand and announce publicly what I had avoided for many years. I cared about, no, I desperately *loved*, this earth. It was time to begin to act on that love. I had journeyed, literally and figuratively, full circle, back to my beginnings, back to the instinctive desires of my early childhood, back to Alaska, back to the risks of caring for a place again, of beginning again with my own daughter, of facing my own self. In another way, also, I had come back to my pacifist roots. For I had decided to quit making war.

The practice of loving, of course, is not an easy thing. It's about letting go. It takes time. Yet, like any journey it begins with the decision to turn and take a step.

I stood there a while longer, facing east. I felt the first outlines of a scary new freedom, an opening to something that I had long yearned for, a release from some long-held confinement. I turned back and began walking to the pickup. Ahead, far off over the coastal plain, a giant plume of black smoke, a flare-off in the oil fields, trailed northward across the sky.

Joan Kane

Due North

I should have my hood on—
Already there are rumors of darkness.
I should see the stones set before me,
Giving passage towards a place
Of complex nostalgias. And now
Should see the scree falling

Endlessly from the mountain's summit,
Falling on the recessive plain.
It is a private place, a wilderness
In practice. I am told that I should look
For a roof in rain, for a river
Split down to tongues of ice.

I shall start all together. As hollow
As a drum, the ground sounds—
It summons, repeats beneath me.
It is as intact and unchangeable
As the seven stars spun into position
When the day, which takes hours to fade,

Has dropped away in its small mist.

Nancy Lord

The Experiment

IT WAS THE FIRST COLD NOVEMBER DAY, with temperatures in the twenties and a bluebird sky to backdrop the snow-covered volcanoes across the inlet. At two o'clock, when I set out for a walk in the woods, the sun was already sinking toward Cook Inlet. Heavy frost from the previous night still coated the trail and crunched underfoot like broken glass. I snapped my jacket closed at my neck, pausing to admire the way the low-level sun lit up the spent fireweed stalks and their cottony strands. *When the fireweed goes to cotton, summer's soon forgotten*. Fireweed has always acted as the Alaskan calendar, the old saying hastening us through summer as the purple blossoms climb their stalks, close into seed sheathes, and—finally—pop open in a white, wind-driven flurry.

The trail I set off on I knew well, although I knew it not as a hiking trail on bare ground but as the start to many miles of winter ski loops. In recent years I've visited the trail system less often than in the past; our snow season has been beginning later and ending earlier—and too often is ruined by midwinter rains. The local newspaper had just pointed out that thirty years ago this same week in November we were digging out from a twenty-eight-inch snowstorm. I recall those snow years fondly—the fences between hayfields disappearing under drifts, the great quiet that settled over the land, skiing until Memorial Day. It may be easy to be nostalgic for the conditions of one's youth, but in fact I now inhabit a different, more temperate climatic zone. Airport weather data supply the evidence: since 1977 mean winter temperatures have increased by six degrees Fahrenheit.

The trail system starts out on state land known as the Homer Demonstration Forest, an area set aside for forestry research and education. One fenced quarter acre shelters experimental plantings of native and non-native trees and shrubs, and another fenced "exclosure" demonstrates what happens when moose are prevented from eating the natural vegetation. (Willows grow big and bushy.) Volunteers maintain the trails and a few wooden boardwalks and bridges.

I stopped in the middle of said forest to watch a pair of eagles and a raven soar in circles above me. If I had been in a true forest, of course, I wouldn't have been able to see the birds past the trees, but the demonstration forest has a problem with its forest. There are trees, yes—spruce trees—but they are bare and broken and fallen, and the sky is large in the holes around them. Forest—with its shade, undergrowth, and damp floor—has given way to sunlight, chest-high grasses, and fireweed. The forest is turning to grassland.

A newly fallen spruce lay broken beside the path, its splintered wood a warm yellow, dry and punky. The tree had not been dead long, and its bark still clung tightly, sticky with globs and drips of pitch.

This has become typical forest for Alaska's Kenai Peninsula—that is, dead spruce trees, needleless and dry, brittle, blowing down in every windstorm. The mature and even the much younger spruce of the region—white, Sitka, and the dominant hybrid known as Lutz—are almost entirely dead now, killed by a spruce bark beetle epidemic that began in the 1980s and swept through four million acres in the 1990s. This death by beetle is said to be the single largest insect kill ever recorded in North America. When I flew over the peninsula in the beetle-spreading years, I looked down on whole sections of forest tinged red—recently killed, with needles turned to rust—and then the next year I saw those sections gone gray and the red spread farther, and then, in a few years, I witnessed the gray trees snapped off and lying over the country like toothpicks tossed down by a giant. Then the logging roads came, cutting up the land and silting the creeks, and trees were mowed down and hauled off as wood chips to become Asian newsprint.

Hiking again, I peered under still-standing trees to where the mossy forest floor was littered with cones and twigs, with oak ferns and species of *Pyrola*, the wintergreens that hold their color through the snow season, the better to get a start on spring. Trees that had fallen across the trail had been bucked up and dragged to the sides. I remembered when the beetle plague first spread and the blame was assigned to poor logging practices. The forestry experts declared that when trees were cleared for homesites and power lines, the downed trees became havens for the rice grain–sized beetles, which reproduced in them and then struck living trees. The solution, those experts said, was to buck the cleared trees into short lengths, because then the wood would dry out and discourage the beetles.

It took years for anyone to understand the severity of the beetle attack, and longer to grasp the cause. Spruce bark beetles are a natural part of our environment, and the records of tree rings and lake pollen show that they have regularly—every fifty years or so—thinned the forest. In our wet climate they are thought to be *the* force—instead of fire—for regenerating forest lands. They play a very useful role in decomposing dead wood. However, nothing in any record suggests they have ever before been so successful in killing off entire forests.

This is the question: what happens when you increase air temperatures by several degrees? I've used the word *you* because I mean *you*, *me*, the *humans* who, through our activities since the beginning of the Industrial Age, have added so much carbon dioxide and other greenhouse gases to the atmosphere that temperatures have warmed by one degree Fahrenheit globally and considerably more toward the earth's poles. The answer in my home place is this: trees accustomed to cooler temperatures and greater moisture become stressed and thus more vulnerable to pests and disease, while bark beetles, previously held in check by cold winters and cool summers, flourish. The beetles are able to complete their life cycle in one year instead of two, and the summers that make us sweat present them with the perfect warmth for mating flights and egg-laying.

At another turn in the trail, I rested my hand on another downed tree, its splintered wood faded to gray. The bark that wasn't already scabbed off was peppered with beetle holes. I pulled away a piece to expose the labyrinthine galleries left by the beetle larvae as they ate their way through the tree's phloem. When a spruce's phloem, between the bark and the wood, is no longer intact enough to move nutrients throughout the tree, the tree essentially starves.

I thought about the summers when the beetles were at their worst, and the hot days that brought them out of the trees to fly in clouds across the land. We brushed them from our clothes and tried not to inhale. We fled the outdoors and made comparisons to Alfred Hitchcock birds and Biblical plagues.

That summer horror has lessened now, only because the beetles finally ran out of food. Without the trees to support them, they faced their own death—or flew to farther forests.

To be sure, not every tree in the demonstration forest is dead. Here and there along my way mid-age spruces still spread their living needles, still sported clusters of cones. And along the trail edges and in new openings that hadn't filled with grass, hundreds of baby spruce trees, like miniature Christmas trees a foot or so high, were springing up. A few among the recruits were tagged with yellow or orange flags, part of a study that involved some plantings and some natural reseeding, in different locations, on nurse logs or in scarified areas where mineral soil was exposed.

One of these flagged areas lay adjacent to an enormous old stump, two feet across, rotten in its center and clothed with lichens. Enough surface was still visible for me to count fifteen or twenty growth rings to an average inch. Ring records don't just tell us the age of trees but show us, in old trees, patterns of tight and wider rings corresponding to a history of forest conditions—moisture and drought, sun and shade. Foresters can typically track the "releases" of greater growth corresponding to periods when neighboring trees died and fell away, decreasing competition for space, light, and nutrients. The old stump could show me a history of bark beetles coming and going through the forest, but I was more impressed by the tightness of the rings generally. Northern forests grow slowly, and that one tree might have been a seedling when Captain Cook gazed upon the spruce-lined shore two and a third centuries ago.

Crows and ravens flapped between trees, and a pair of magpies—*yak, yak, yak*—followed one after the other through an opening. Insect-eaters—chickadees and woodpeckers—have done well in the dying and dead forest, and the omnivorous magpies seem to be everywhere. Magpies favor disturbed areas, including open woods and fields, and they are significant predators of other birds' eggs and young. According to bird-sighting records, their numbers have increased mightily throughout southcentral Alaska *and* they're moving northward, deeper into Alaska's interior where they have not previously been players.

I came finally to a more open area, where a wooden boardwalk crossed wetlands and grass gave way to willows and crowberry. At the edge of Diamond Creek I listened to running water and admired the lacey configurations of newly forming ice. Diamond Creek is just one of the many waterways in the region that, due to record-setting precipitation and unusually warm temperatures, suffered not one but two "one-hundred-year floods" in 2002. The flooding severely altered habitat both in the stream itself and along and beside its banks. Beaver dams washed away, and the beavers themselves disappeared, presumably drowned. The Dolly Varden that lived in Diamond Creek have also disappeared, their streambed habitat scoured clean.

Diamond Creek does not support salmon, a mainstay of Alaska's economy, but neighboring streams that do are threatened by warming waters. According to Alaska's water-quality standards, 55.5 degrees Fahrenheit is the upper limit for successful salmon spawning and egg survival. On the nearby Anchor River that temperature was exceeded on eighty-eight days in 2005. Moreover, temperatures of sixty-eight degrees, the assumed limits of salmon tolerance, were exceeded on six days. Biologists and fishermen were monitoring salmon health and production nervously, well aware of the situation farther west, where Yukon River king salmon have been devastated by a fungus associated with warmer water.

What happens when you increase the earth's temperatures? Warming rivers, hotter and longer summers, warmer and shorter winters, thawing permafrost, melting glaciers, thinning sea ice, drying wetlands, shrinking lakes, more wildfires, species migrating into new ranges, species facing extinction, disrupted breeding schedules, declines in food production, lighter snowpacks, eroding coastlines, displaced people, dead forests, temperature-related diseases, food stress, rising sea levels, ocean acidification, landscape transformations, stormier weather—here in the north we've already got all this.

Do you think what happens in Alaska doesn't and won't affect you? Think again. This is your future. "Business as usual" will cause the global mean temperature to increase by between 2.5 and 10 degrees Fahrenheit by 2100, according to the scientific consensus report prepared by the Intergovernmental Panel on Climate Change (IPCC), an international collaboration of thousands of scientists recognized as the world's leading authority on climate change. The results of such rapid temperature increases, the fastest in at least ten thousand years, will be dramatic in a world already straining to provide food, water, and shelter to 6.5 billion people.

The question we're now testing might be: once a species evolves to be capable of actually changing the world's climate, can that species manage and modify its behavior to prevent the destruction of the systems that sustain it? Or will it be the agent of a mass extinction? In such an extinction, what will be the fate of that apex species? On our singular, beautiful, and ravaged earth, this is an experiment without controls.

As I circled back to the trailhead, I thought about that species—so far evolved, so able, so intelligent. I thought about the clock analogy: if a twenty-four-hour clock started at the earth's beginning 4.6 billion years ago, it was already 7:15 p.m. when the first multi-cell plants evolved and 9:25 p.m. when the first fish swam in the oceans. The first small African apes didn't appear until 11:58 and *Homo sapiens* himself until 11:59:58—two seconds ago. Perhaps we should wonder if we aren't the experiment, like a small test light turned on before blinking out, forever.

John Morgan

The Moving Out

After sunset when the grieving
move further into their grief
and the stars are revealed by their master, the darkness,
I have left the cities of the blind
along tracks straight and cold as the north.

Here I sit listening on the shore
of a white and glacial distance.
The voice of a girl like an opening flower
begins to curl forth from the inner shell of the mind.
So many nights I have waited.

In cities the darkness gobbled me up and spat me out,
my fears scuttled back and forth outside the door.
Now the first birds waken and peck among fresh snow.
The light begins to open
with a pink and icy whisper along her cheek.

Karin Dahl

Going for Water

WHEN I WAS A CHILD I lived with my parents in a green army Quonset without plumbing or electricity. We were homesteading forty acres of land in Willow. The previous owner had tried to dig a well on the land, but because he didn't dig deep enough, or for some other reason, the water was never clean enough for drinking or washing. So from early summer to late autumn, every few weeks my parents took me in our old blue Ford pickup to the spring to get water.

In winter, when the snow piled high, it was difficult or impossible to get to the spring. All winter long we got our water from the faucet at the Willow store. But during the summer we went to the spring, where the water was fresh and ice cold, and where it didn't cost anything more than the work to go and get it.

I sat between parents as my father drove about four miles south on the Parks Highway from our homestead. There was no road down to the spring, so my father had to turn the truck crossways off the highway and back down a sharp incline over low brush, weeds, and loose gravel. I was always frightened then, feeling like we could roll over and down the hill as easily as not.

Once he had backed the truck down over the hill and a short way to the spring, we reached an open area. It always seemed to me a magical place. The hill that we had come down sheltered us on one side; on the other side a hill led up to railroad tracks. Behind us were delicate birches, alder, and spruce. We rested in a small clear space, next to the stream that the water formed as it poured from the ground. When the truck's engine was silent, all we heard was the soft sound of the water, an occasional birdcall, and our own voices.

While my parents took old Clorox jugs to the spring to fill them, I played in the cold stream or climbed the hill to the railroad tracks. There I could walk on the rails as if on a balance beam, or dig in the gravel for something that might have fallen off a train and been lost there. The bitter, dark aroma of creosote-coated tiles filled the air. In early summer it blended with the sweet smell of new plants and tree buds, and in autumn with the tart scent of cranberries that weighed down bushes and piled up on the ground.

From where I stood on the tracks, I could see how the land spread out and away beyond us. If I looked east, there were no houses or roads, only a dense tapestry of small trees and bushes that stretched to the horizon. Without words for it, I had a sense of the land's harsh delicacy, its fragile strength. In the cold, slanted light of a late afternoon, when the sky was washed pale, that landscape could fill me with a piercing sadness. It was the harsh, unrelenting beauty of the land itself that made me feel that way, and it was also the sadness and beauty of the life I shared there with my parents.

Once they had filled the jugs and lined them up in the bed of the truck, my father and mother sat in the truck's cab with the air of weary companionship that by this time characterized their relationship. My father's tall, thin frame bent over the steering wheel as he smoked a cigarette. His face bore a gentle, bemused expression. My mother put her crutches back behind the seat. She had used them ever since she had polio several years after she and my father were married. The doctors had told her that she would be lucky to learn to sit up again, but they didn't count on the iron determination that lay beneath her lovely features and quiet manner. Only an occasional glint in her green eyes, a sudden directness of her gaze, hinted at her unflinching will. She was the one who had arranged for us to take over the Willow homestead. She had grown up south of Willow, in the Matanuska Valley.

Our simple life out on the land was supposed to save them, to save my father from the melancholy and drinking that were consuming him, to save my mother from the anger and despair that were leading her to drink as well.

That did not happen.

But the harsh, broad beauty of the land filled each of us. We listened to the long cry of loons as they swam on the lake down from our Quonset, to the hoot of owls in the trees around us, to the small song of chickadees. We watched as the light waned into the darkness and deep snow of winter, and then as the light returned with spring. My mother taught me how to sail small birch-bark boats down the streams the melting snow made on the long road to our house. My father tapped the birch trees so that I could taste the sap they lived by. And again and again, all through the summers, we drove down to the spring for fresh water.

Pamela A. Miller

Beached

Black and white
Seabird on sand

You run a short dark way
From my reaching hand

And stop, rocking
Side to side on sternum.

Grasping your shoulders
I check for broken legs

Torn wings or sticky oil.
Only suspiciously thin breast.

Maybe you are just an old murre.
Perhaps you got lost riding in fog.

Running to the basement of a wave
I give you a gentle toss.

Like Styrofoam you float
But do not dive or fly.

Rolling out with murky foam
I carry you back to dry beach.

I should get some herring
Or warm you in my down coat.

I could throw you farther out to sea
Collect as sample, or break your neck.

Gull strides beside you,
Ruffles clean feathers.

Crow eases low over
Your last heavy blinking.

Molly Lou Freeman

Drink

Hard up (on a reef) the skipper wept (into his bunk and his bottle)
while she—craftily—banged (sat tight)(shook)

a little—someone radioed in
looks like we're screwed
to a rock.

In the prelude, the iodized air smelt frond-
like, like iron
cold suckle and gut

fresh—when she was yet an April still thing—
—laid so deep in the nightfall.

When despair was distant—
that was *before anything, anything bad had*
happened—some boats with booms came.

Someone rode a bicycle down the deck and sniffed out
the Sound
smelt then

the hull hole,
perfume of earth's ass, earth's anus
which lubed, spewed-slow-ribboned the sea's self.

Contagion at daybreak,
not blue—shone—she
looked silked, looked sauced.

Who would wash the tarry feathers—
& how would they
die there, these beasts?—

So fully licked.
You wouldn't want to put into that
our sweet bottle of Joy.
What a sad catch

water haul and tar-
balls beneath a star-
strewn passage.

A chopper
hugging the shore,
groped. She
in her dress,

she rode coastwise and in
to Afognak and Dutch
went leagues beyond the mind's

measure—the sad fisherman
on his boat he too weeps—my skipper.
The long lines, the net not

unwound, worked—she went crudely on and in,
I said she wore black
which sunk and stank beyond law

my dark girl, my dark gulf well-strapped
& stacks barren
of birds.

Hello
All

Vessels.
The mind is wider than—
the Sky.

The mind is deeper than—
the Sea
says my log.

A malady—we await money.
Ask which beasts are
all gone and for how long
—if one oyster-
catcher garnet trills.

Still we highline, run her on the high
tide,
her beauty misleading—

her treasure.
We—full fathom
only little—retell:

The Sea, Love, the Deep
Drink
the Mind and its Capacity

& God & We

[buy rounds]
[sing a cheap song]
[we don't go *down with this ship.]*

Walter Meganack, Sr.

The Day the Water Died

GOOD DAY, LADIES AND GENTLEMEN. I want to thank Mayor Devons for arranging this historic international event, and for inviting me to speak on behalf of the villages of the Chugach Native Region of Alaska.

The Native story is different from the White man's story of oil devastation. It is different because our lives are different; what we value is different; how we see the water and the land, the plants and the animals is different. What White men do for sport and recreation and money, we do for life: for the life of our bodies, for the life of our spirits, and for the life of our ancient culture. Fishing and hunting and gathering are the rhythms of our tradition, regular daily life times, not vacation times, not employment times.

Our lives are rooted in the seasons of God's creation. Since time immemorial, the lives of the Native peoples have harmonized with the rhythm and the cycles of nature. We are a part of nature. We don't need a calendar or a clock to tell us what time it is. The misty green of new buds on trees tells us, the birds returning from their winter vacation tell us, the daylight tells us.

When the days get longer, we get ready. Boots and boats and nets and gear are prepared for the fishing time. The winter beaches are not lonely anymore, because our children and our grown-ups visit the beaches in the springtime and they gather the abundance of the sea: the shellfish, the snails, the chitons. When the first salmon is caught, whole villages are excited. It is an annual ritual of mouthwatering delight. The children are excited, the parents are pleased and proud, the elders smile in their memories of seventy-three other springtimes in the village.

When our bellies are filled with the fresh new life, then we put up the food for the winter. We dry and smoke and can. Hundreds of fish to feed a family. The homes have hanging fish alongside hanging laundry. The sights and smells of a village in the spring. This is the Native way. This is the way the elders taught us, and their elders taught them, for thousands of years. Thousands of years. Since the big ice left Alaska. There was no Europe then. No Roman Empire. There were no Jews, no Christians. No Egyptian civilization. But my people were here, the Alaska Natives were here, celebrating spring and life and laughing and loving and working and teaching. The rhythm of nature. The rhythm of our lives.

Much has happened to our people in recent centuries. We have toilets now, and schools. We have clocks and calendars in our homes. Some of us go to an office in the morning. The children go to school in the morning. But sometimes the office is empty and locked. Sometimes the child is absent from school. Because there are more important things to do. Like walking the beaches. Collecting the chitons. Watching for the fish.

The roots of our lives grow deep into the water and the land. That is who we are. We are like our brothers the bear and the deer—we live on the land, and our food is mostly from the water. Bear eat fish, deer eat seaweed, Natives eat all of the life in the water.

The land and the water are our sources of life. The water is sacred. The water is like a baptismal font, and its abundance is the holy communion of our lives.

Of all the things that we have lost since non-Natives came to our land, we have never lost our connection with the water. The water is our source of life. So long as the water is alive, the Chugach Natives are alive.

It was early in the springtime. No fish yet. No snails yet. But the signs were with us. The green was starting. Some birds were flying and singing. The excitement of the season had just begun.

And then we heard the news. Oil in the water. Lots of oil. Killing lots of water. It is too shocking to understand. Never in the millennium of our tradition have we thought it possible for the water to die. But it is true.

We walk our beaches. But the snails and the barnacles and the chitons are falling off the rocks. Dead. Dead water. We caught our first fish, the annual first fish, the traditional delight of all—but it got sent to the state to be tested for oil. No first fish this year. We walk our beaches. But instead of gathering life, we gather death. Dead birds. Dead otters. Dead seaweed.

Before we have a chance to hold each other and share our tears, our sorrow, our loss, we suffer yet another devastation. We are invaded by the oil company.

Offering jobs. High pay. Lots of money. We are in shock. We need to clean the oil, get it out of our water, bring death back to life. We are intoxicated with desperation. We don't have a choice but to take what is offered. So we take the jobs, we take the orders, we take the disruption. We participate in the senseless busywork.

We start fighting. We lose trust for each other. We lose control of our daily life. Everybody pushing everyone. We Native people aren't used to being bossed around. We don't like it. But now our own people are pointing fingers at us. Everyone wants to be boss, we are not working like a team.

We lose control of our village. The preschool meets in the community center. We shut down preschool so the oil company can have the center. We work for the oil company now. We work for money now. The springtime season of our village ways is gone. Destroyed.

Our people get sick. Elders and children in the village. Workers on the beaches. Lots of sickness this year. Stomach sickness. Head pains. Bad colds.

We hardly talk to each other anymore. Everybody is touchy. Everybody is ready to jump you and blame you. People are angry. And afraid. Afraid, and confused. Our elders feel helpless. They cannot work on cleanup, they cannot do all the activities of gathering food and preparing for winter. And most of all, they cannot teach the young ones the Native way. How will the children learn the values and the ways if the water is dead? Very afraid. If the water is dead, maybe we are dead—our heritage, our tradition, our ways of life and living and relating to nature and to each other.

The oil companies lied about preventing a spill. Now they lie about the cleanup. Our people know what happens on the beaches. Spend all day cleaning one huge rock, and the tide comes in and covers it with oil again. Spend a week wiping and spraying the surface, but pick up a rock and there's four inches of oil underneath. Our people know the water and the beaches. But they get told what to do by ignorant people who should be asking, not telling.

We fight a rich and powerful giant, the oil industry, while at the same time we take orders and a paycheck from it. We are torn in half.

Will it end? After five years, maybe we will see some springtime water life again. But will the water and the beaches see us? What will happen to our lives in the next five years? What will happen this fall, when the cleanup stops and the money stops? We have lived through much devastation. Our villages were almost destroyed by chicken pox and tuberculosis. We fight the battles of alcohol and drugs and abuse. And we survive. A wise White man once said, "Where there is life, there is hope." And that is true.

But what we see now is death. Death—not of each other, but of the source of life, the water.

We will need much help, much listening in order to live through the long, barren season of dead water, a longer winter than ever before.

I am an elder. I am chief. I will not lose hope. And I will help my people. We have never lived through this kind of death. But we have lived through lots of other kinds of death. We will learn from the past, we will learn from each other, and we will live. The water is dead. But we are alive. And where there is life, there is hope.

Thank you for listening to the Native story. God bless you.

This essay is adapted from an address to the Oiled Mayors of France and Alaska at a conference in Kodiak, Alaska, in June 1989, just three months after the Exxon Valdez *ran aground on Bligh Reef. Chief Meganack was asked to speak on behalf of Alaska Natives from the Chugach Region. This public testimony was subsequently reprinted in numerous places, and his words, "the day the water died," have since been used to refer to the oil spill's impacts in books, movies, and articles.*

Buffy McKay

October

I am practicing my regret, now
as the dried-gourd ground of fall
crisps under my feet.
I walk in a line along the drive
where the bulbs were supposed to be
placed in the undug ground
with all the hot optimism
of fresh-sprung love.

After working my swivel chair
desktop day, my spine creaks
like the door of the freezer swung wide
to reveal ice
and old corn
but no fish and no meat
and no berries.
It is modern art in there,
stark, spare, and bare.
No days out in the bogs
high-stepping like moose
to grab salmonberries bright like both
morning and evening,
gold tinged with scarlet.
No blueberries baptized by dew,
no wet butt and knees from
losing my balance from
laughing from
trying not to spill my bucket.

No camp trips or dip nets
or small-rock spinal bruises.
No swimming, no clamming,
no briny pints of vivisected bivalves
dug under starshine extra-low tides.
Not one footstep on Clam Gulch,
not one shovel-stroke underneath it.

No fish eyes to watch me
from chill-stilled heads
lined up in their freezer bags.
No rich red meat waiting
patient and loving
like all good steaks should.
No half dried and half smoked
dunghnak like my mom said,
ready to dip into salt and new seal oil.

No mom to say *dunghnak* properly for me.

Spring carried me into a land
where people rise early and
travel empty-handed,
but for paper.
My modern-art freezer
with its accusing, baleful paucity
sighs me a secret in condensation
that shies away from my breath:
the only way to pay
for time is with regret, doling out
coins of sorrow to the dirge of the keyboard.

I am sorry, I am sorry,
for watching, not touching
the seasons.

Marybeth Holleman

What Happens When Polar Bears Leave

As we stood in line at the grocery store, with our bread, lettuce, and chocolate-chip cookies, this is what my husband told me that made me turn away so quickly I was left dizzy:

His friend Chuck, a government biologist, was flying over the Arctic Ocean for the annual bowhead whale count. While scanning for whales, he saw polar bears, but in ways never before recorded. He saw females with cubs swimming in the sea with no ice pack in sight. He saw a drowned polar bear floating in a sharp blue sea, then another drowned bear, and another, no ice or land, just open water for more than fifty miles in every direction.

Not until it was our turn in line did I turn back around, wiping tears, and focusing on the face of my son. He looked at me, and grabbed my hand in his.

It was my own fault Rick told me when he did. Standing there in Anchorage, seven hundred miles from the Arctic Ocean, I had pointed to the magazine rack, to the cover of *Alaska* magazine: a full-frame shot of a polar bear face. In the corner, over an erect ear dusted with snow, were the words "Bound for extinction?"

Polar bears are excellent swimmers. Though males often weigh more than a thousand pounds, they can swim for miles, their broad front paws like paddles, their long necks and narrow heads reaching above ice-studded water. Polar bears are so aligned with water that they are classified as marine mammals; *Ursus maritimus* means "sea bear."

But they're built only to swim from one ice sheet to another. They are not made of blubber and fin like whales, or seals, or walrus; they are made of fur and foot; they need the polar ice pack.

They are ice bears.

Three years after that day in the grocery store, everyone knows that polar bears are drowning. So, when my husband tells me that, on his last fly-over, Chuck saw dozens of polar bears stranded on a single drifting iceberg, I'm not surprised. When he says there was no ice pack or land in sight, and the bears would all most likely drown, I feel once more that abiding dismay, that upwelling of anxiety, and say to him, "I don't know what to do with that information."

"Yes," he replies, "lots of people feel that way."

In the last forty years, summer arctic sea ice has shrunk by half. In the next twenty years, it may disappear completely. We may lose it all. The polar bears may lose it all.

As the ice pack thins and recedes, polar bears are forced onto land, paddle-shaped paws lumbering across tundra, where they must hunt less nutritious food, where they compete with brown bears who are more adept at terrestrial living.

Or they are left to swim and swim, in search of the ice that was once there, swim to exhaustion, to drowning.

I tell my friend Karin about the drowning polar bears; we talk about dying forests and rising seas and world leaders who do not take action. I am on fire; I want to know what to do. Her voice is even and soft as she both agrees and disagrees.

A Tibetan Buddhist, a high school English teacher, and the mother of a teenage boy whom I have known since his birth, Karin is a compassionate woman; she teaches attentively, but doesn't expect the teenagers in her care to even have a chance to change the world. It will be too late, she tells me, for them to grow up and vote new leaders into office, and for these new leaders to stop the rising seas and mass extinctions.

But her own son, her one-and-only, lifelong friend of my son James, what is her hope for his life?

"That he become a monk," she says quietly. That he pray and transcend this world that is soon to desert him. This world, according to her religion, is merely an illusion created by our wanting minds.

"Of course," she tells me, "it's not that simple."

She plants lilacs in her yard and tends their new growth; she creates her own hospice care for a dying cat. But she is already loosening her hold on the world even as she gets up every morning, sends her son off to school, enters a classroom and asks her doomed students to care about getting an A in English, to write a sonnet, to consider what an eighteenth-century novelist had to say about life.

Already some areas in the Arctic are warming ten times faster than the rest of the planet.

Already arctic sea ice forms later in fall, depriving coastal villages of the natural barrier from fierce storms, which now erode their shores and flood their houses. Of 213 Alaska Native villages, 184 face flooding and erosion.

Already the sea surges and rises, permafrost melts beneath buildings, and the people of Newtok, Kivalina, and Shishmaref have begun plans to move their villages inland—refugees of global warming.

Already the ice breaks up weeks earlier in spring, constricting the time villagers have to hunt walrus and oogruk. Constricting hunting for polar bears, too; the earlier the breakup, the poorer the condition of the polar bears. Declining conditions are most pronounced in their southern range: the bears are becoming thinner; females are giving birth to fewer young; cubs are taking longer to wean; fewer cubs are surviving to adulthood. In Hudson's Bay, for every week that the ice breaks up earlier, polar bears come ashore twenty pounds lighter. Within a decade, females could become so small that they won't be able to bear young.

The polar bears shrink as their ice shrinks.

Driving to pick up my son one August afternoon, I had to turn on fog lights to see my way. But there was no fog, and the sky was cloudless. It was the third day of smoke so thick it hid the mountains around me, coarsened the throat, kept children inside and headlights on all day.

That summer of 2004 was the worst season of wildfires in Alaska's recorded history. Over 6.4 million acres of forest went up in flames. The temperatures were, for the sixth time that month, at a record-breaking high. The leaves of birch trees, which should have been a late-summer deep green, were brown and shriveled from a record drought, green gone a month early, fall's yellow skipped.

Record wildfires. Record drought. Hottest summer ever. Anchorage shrouded in smoke, the broad peninsula of roads and neighborhoods, downtown skyscrapers, unruly braids of lakes and parks and greenbelts, all cloaked. The snow-spiked peaks of the Alaska Range across the broad expanse of Cook Inlet, all the simmering volcanoes, gone from view. Even the familiar lines of the Chugach Range edging the city disappeared.

Enshrouded we sat in our cars, driving our SUVs here and there as if there was no connection between our driving behavior, our big-car boom, our life in cars, and the smoke, the fires, the extreme heat and drought. As if we could drive and drive, lumbering beasts unable to see disaster in time to change direction, drive and drive, and the skies would clear, the sun hanging just so above us, the world around us immune to the unchecked spread of our daily habit.

And I, who knew the truth of it, I, along with the rest, driving.

Writer and teacher Carol Bly says she most admires writers who can speak of beauty and horror together; who can describe a shimmering landscape, and then have some dark thing happen in that landscape; who can write of Eden, and the fall of Adam and Eve.

It's an old story, beauty and horror, and we're drawn to it, drawn to just a handful of tales, told again and again with different details, same endings. Stories of forgiveness, rediscovery, redemption; of joy, wonder, light emanating from—*surprise!*—dark, sorrow, despair.

Would I have remembered what my husband had told me if the arctic ice held no allure? If the polar bears had just been beautiful, and not also doomed? If his friend had seen their shimmering whiteness against the glazed blue of multiyear ice, and not small cubs swimming with no ice in sight, not the body of a white bear, face down, in an azure sea?

Would it have mattered?

I was thirteen on the first official Earth Day. Same age as my boy now. After school, I walked the neighborhood alone, thinking of the planet and of my adult life before me. It was the first time I'd thought of the Earth as a living entity, as something I could affect. I scanned the sidewalks and roadsides, looking for litter. I picked up one soda can beside the road, all the litter I found that day. Just one, but I still feel the coolness of that thin empty container, see it glimmer in the afternoon sun, still savor the heart-skipping lightness I felt the rest of the day.

What was reflected in that soda can—I wanted more of that feeling. I wanted to be of use.

But one soda can is nothing. Did no good. What does?

Ringed seals make up ninety percent of a polar bear's diet. Ninety percent.

In April 2002, the bodies of hundreds of drowned seal pups were found in the Gulf of St. Lawrence. The sea ice melted earlier, the ice floes disappeared sooner, and the mothers, who congregate on ice floes to give birth, had been forced to bear their pups in open water.

Global.

Warming.

I can't make sense of it. I can't read about it. I can't talk about it. It's too big. What—large, stronger, more frequent hurricanes in the south, more rampant fires in the north? Another shelf of the Antarctic fallen off, adrift, melting? Animal populations shifting northward, northern species like polar bears and lynx running out of room? What can I do about all that? Here—I'll recycle newspapers. I'll pick up litter in the park. I'll bring my own coffee cup to the latte stand. I'll even think about driving a smaller car, or a hybrid. Promise. OK? Good enough?

Face down. An azure sea. No. Not good enough. None of this. None of the two thousand things I can think of to do is enough.

I resolve to get up earlier every morning and meditate.

I resolve to practice loving kindness, to be generous and compassionate toward all things.

I resolve to want what I have, and no more.

To be grateful, and aware.

As I sit in the doctor's office, waiting to begin the treatment to rid my body of abnormal cells, to ward off cervical cancer, I take long, deep breaths and try to think good thoughts.

As I drive to pick up James from school, I count my blessings, even as rain falls midwinter and melts the banks of white snow, the weather itself a grim reminder.

My doctor has given me a prescription for Flexeril to help with tension headaches. Take one or two at night, he says. One or two, you can take them indefinitely, no side effects, no addiction, no worries.

Look, I tell my husband, I can refill this three times. I like the way they help me relax, help me sleep. They're so nice for long airplane rides, especially if the plane is late or cancelled.

But the headaches do not go away. They do not even lessen. The pain spikes when I laugh, or roughhouse with my son, or run with my dogs.

I am also supposed to do neck traction every day, but I forget. I have a hard time doing anything for myself when the world around me burns.

My husband, a university professor who specializes in conservation, boards a plane at midnight, and flies halfway around the world to help a distant impoverished chaotic country deal with environmental damage caused by oil. The irony of the jet fuel it takes to get him there does not escape him. We laugh as I say he's off to save the world, but beneath the laughter we're dead serious. He is driven to try, and it's what I love most about him. After he leaves, though, I wonder how much of our striving is simply a desire to *do*, to *have done*—that restlessness we feel when faced with despair, the way constant movement clouds truth, defies meaning. It's as if we, too, are swimming, swimming, toward a shore we cannot see but still believe is there, somewhere.

Why did I make my son get up and go to school an hour early for jazz-band practice? Why do I keep making James try things, track or band or honor society?

He is a bright boy, *gifted* they call it now. But he does not want to go to school. He does not want to play trombone or basketball. He only wants to lose himself in computer games, pretend to be something/somebody/somewhere else.

Why does this confound me? Do I think he is immune?

If the world is burning around us, and there's nothing each of us can do, if all that's left is to pray for one's own vision to see past this illusion, then why in God's holy name would I push my son to do all these other things?

What is it like for him, for all the children growing up now? When the American Dream of abundance and opportunity is so frayed and worn that, really, who believes it anymore?

They say that generation Y, the label for my son's generation, are homebodies. These kids like to be connected to family and friends. They like computer chat rooms and cell phones. They work collaboratively and do not like to be alone. Is this a form of busyness, of denial? Or is it some precursor to a solution, water on the flames?

I wonder what we'll call the generation after generation Z. I wonder if we'll start over at A, or if we're not expecting to be around that long.

A picture on my desk shows my son at one year old, walking in the shallow end of a clear lake on a mountainside in Prince William Sound. He wears nothing but his diapers and a white T-shirt. His blond curls have not yet been cut, and they drape down over his shoulders, a cascade of light. The August day is sunny-bright, the lake's water is clear enough to see the stones at the bottom, the edges are laced in wildflowers and wind-sculpted spruce. But the ripples he makes as he steps through the water break apart his own reflection.

The polar bears' images awaken me in the middle of the night. Sow and cub, swimming in circles, no ice, no rest, in sight. Scores of white bears crowded together, stranded on a melting chunk of ice.

An increase in spring rains is causing their snow dens to collapse, killing females and cubs; earlier spring breakup separates dens from feeding sites, farther than cubs can swim; starving adult males are resorting to eating the smaller females, cubs, and yearlings.

This cannibalism, says a local environmental activist in one of a growing cascade of newspaper stories on polar bears, is "global warming's bloody fingerprints."

Mothers and cubs first.

I've never seen a polar bear in the wild. I want to go to Churchill, or Barrow, I tell my husband. We have to go see them before they're gone, I say, frantic. I can't stand the idea of not ever seeing them for myself, even if it's at a garbage dump outside a town.

But worse, I cannot stand the idea of living in a world without them. I may not have seen them in the wild yet, but I've seen them. We all have—their images are everywhere. On every nature calendar's December or January. On Christmas cards. On commercials for soft drinks. On hockey teams' emblems. Everywhere, their faces, that whiteness punctuated by two dark eyes, the dark nose. We are enamored of them.

I remember learning, soon after I moved to Alaska two decades ago, that polar bear hairs are hollow, and it's the light filling the shaft that makes them look white—a reflection of snow. Those hollow hairs help regulate temperatures, acting as heat escape vents. Polar bears overheat at temperatures above fifty degrees Fahrenheit.

I remember learning about how they crouch on ice by a seal's breathing hole and wait. How, according to Native lore, they hide their black nose with a paw, or clump of snow, making the disappearing act complete.

My son tells us one night, after dinner, that polar bear skin is black. This is something my husband and I didn't know. We look it up and James is right: their black skin captures heat while the hollow hairs release it.

Even if they do become extinct, which seems not just likely, but inevitable, we will still have the pictures. We'll still have that image of the white triangle of a cub sitting between the two front legs of the mother, sturdy pillars of protection. And the bear lying on his back, hind legs splayed in relaxation, paddle-shaped front paws limp and resting by his furred face.

Extinct, would they still appear in calendars? Would we still see their likeness in cartoons? Or would we banish their image as they have been banished? Would guilt and grief win out, or would aesthetic desire for beauty rule?

I've never seen images of other recently extincts—sea cow, passenger pigeon, dusky seaside sparrow—casually displayed in popular culture. But in this age of accelerated extinction, that might change.

Suppose an animal's extinction was rapidly followed by the erasure of their image from our grasp. Suppose we had to remember them only in our minds, our hearts. Suppose our struggle to maintain that fading image in our mind's eye was ultimately futile, as it is with the face of a lost beloved, always receding, always disappearing, no matter how much we want to hold on to it.

What I want to know is, what will it take for us to save the polar bears? What will it take for us to turn this around in time for polar bears, and ringed seals, and walrus? For generations Y, and Z, and A?

Tell me and I will do it.

The lord of lords, the one we all fall down on our knees to at one point or another in our small sweet lives, even she will not tell me. Supplication provides no relief, no answers. There is no solace for those who do not avert their eyes. And I will not avert my eyes. And I beg you to not avert your eyes.

If I were to ask you, if I were to ask everyone I meet, do you want polar bears to become extinct? To a woman, to a man, you'd say no.

I have so many questions without answers. I fall down into the darkness of them. I have only images, handfuls I grasp and sift through, looking for answers by a light so dim it could be coming down a shaft of hollow fur.

There's a boy, sleeping. His mother tells him a story each night at bedtime. She has told him a different one, one she creates as she speaks, every night of his life. As he grows, she worries the stories are too simplistic, and should become more realistic, like the books he is reading, where everything does not end happily and animals don't talk. But these new stories don't help him fall asleep. Instead, he tosses and turns for hours. He has nightmares, and stomachaches in the mornings before school. So she begins incanting the simple stories again. She returns to the world inhabited by animals who talk and learn and love the way she loves the boy, the way he loves her. And every night he lies still and listens; he says, "Good story"; he falls asleep. Each night she describes a world that slips away each day. More and more her stories are lies, and she knows it, yet still one comes to her, every night, and lulls them both to sleep.

"Look twice: once for love, once for survival."

Patricia Hampl has faith, she says, not in what will or won't happen, but in lyricism, "an authentic response to the world's impossible contradictions which seem to resolve themselves, finally, as beauty." It is the artist's work not to celebrate, she says, but to express wonder.

"And something terrible resides at the heart of wonder."

Golden birch leaves lie scattered over outstretched boughs of spruce, creating a forest of Christmas trees three months early. I point them out to my son.

"Yeah, pretty," James says.

It was difficult to pull him away from his new Sims computer game, and at first he trudged along, head down, angry with me. Now he watches our dogs stop to sniff a clump of Canadian dogwood as if the red leaves held some new story.

"I wonder what they smell," he says, "it's like they have this whole other way of knowing things."

I tell him of research about dogs who know when their owners are coming home—not just by time of day or hearing the car, but from miles away, as if they have another sense.

"Oh, yeah, of course they do," says James.

We climb the wooded hill, golden leaves drifting down around us, and my gait lightens.

"Even knowing that the horrible and beautiful are together in the world, we pass the threshold into something finer," says Linda Hogan. It's the very act of recognizing, through all the layers of one's being, that the horrible exists side-by-side with the beautiful, and that it's in this coexistence, this rubbing against each other, that life glows brightest. A fine and lovely line.

Wait. Don't drift off: we're really going to lose polar bears, all of them, in a lifetime.

Listen: since the drownings began, new seismic exploration and oil drilling, onshore and offshore of the Arctic Ocean, has reached an all-time high. Right now, as their ice shrinks, we are at fever pitch to pull the very substance out of their ocean home that we will burn into the very substance that is destroying their ocean home.

I can't stop yet. I aim to write into an answer. Or at least a beginning, the dim outline of a shore. Can you see a shore? If I stop swimming, I die. So do you.

Fear-mongering. That's what Alaska's only congressman called reports of global warming. "I don't believe it is our fault," said Don Young. "That's an opinion. It's as sound as any scientist's." It's just natural causes, he said, like an ice age. And in the next breath he said, even if it is human-caused, "You're not going to turn it around unless everybody stops living."

After public testimony on the effects of warming, a member of the state of Alaska's Climate Change Commission asked, "What about human breath? It's a major cause of warming. Look it up on the web." He was serious. As was another member who said, "Well, we need to remember the many positive effects of global warming." It'll be a great boon to fisheries, she said, it'll be good for Alaska's economy.

Increasingly, when I talk with others about polar bears, they say, so quickly and with such serenity, "Oh, polar bears are toast, there's nothing we can do for them now."

How is it that we can move so seamlessly from denial to fatalism?

Not by flood, said the Lord, will I end the world next time. Not by flood. Maybe it will be by fire, then, heat rising, global warming making the continent of Australia so hot that parts of it, scientists predict, will soon be "like living in an oven."

What has protected polar bears up until now, states one report, has been their isolation, the relative lack of human presence in their sea ice world.

As it becomes more difficult to find and reach seals, polar bears are seen more often, and in larger numbers, near villages. They are especially drawn to whale carcasses left from fall subsistence hunts. In Kaktovik, on Alaska's Arctic coast, polar bears start appearing in early September, waiting for the whaling crews to begin.

Polar bears are shrinking, vanishing. They are desperate, and moving closer to us with each desperate step.

"Nothing in the world can take the place of persistence," claimed Calvin Coolidge. His world was such a far cry from ours today, though. Could persistence save polar bears, the ice pack, and generations yet to come?

Horror and beauty; wonder and terror; black skin and hollow fur.

It is a dark November evening. The snow came early this winter, then melted. Now ice runs up and down the roads, up and down the trails.

James and I took the dogs for a walk yesterday, through spruce and birch to Goose Lake. The low sun shot light across the white expanse. A dusting of frost over old snow glimmered into the sky's deep blue. The dogs' gait loosened, from the nose-down straight-on intensity of woods walking to an open jaunt. The breadth of ice and light sent all four of us sprinting and sliding over the frozen lake.

James laughed. "Oh," he said. "I love this ice. I love winter."

Now in this dark I remember sunlight bouncing off ice, and I call my son to bed. It's a school night; morning will come too soon and too dark, and it will be a struggle, but we will open our eyes, we will arise, we will skate through the difficult, the uncertain, the lovely day.

Amy Crawford

Bones

1.

I am a thing sculpted by footfall
day after day, over rocks and tundra,
along game trails or no trails on high passes.

I cross over bear tracks laid in sand,
just formed, nearly warm. We each pass
our ways privately. In my tent

I read, write, invent sense out of this life,
humming words into lines: words,
raining thoughts, water for my landscape.

2.

Walking coral beds smooth as marble,
beds with million-year fingers like sage,
with small round crinoids, with coral

apartments from equatorial seas . . .
teracorals and streptalasma . . . brassy
shales, green and purple slates crunching

beneath boots, rising in vertical columns . . .
limestone arcs offset by a fault, the sweet
cusp of chevron folds . . . these are the bones,

the old weathered sediments, the baked
and up-thrust, cracked and twisted,
torn and broken-down bones of the land.

3.

Sometimes I am visited.
After the wind spends days blow-
drying the sky, no breath left,
the valley lies stark naked of sound.
I lie at night under the giant starless silence
listening to flower petals curl to sleep
like wolf tails, the whisper of raven
feathers; listening for the breath
of a bear, which does

come, if you travel for a time in the north.
Usually, we are equally startled;
I holler *hey hey hey* and the bear
grunts and thunders off.
I crawl from my tent and stand naked
so as to see the maker of sounds.

4.

The riverbed stands empty of
sound, its stone trenches
waterless, like a silent idea.

That is, without minds, ideas vanish;
without rain, the drainage has become
a bed of old stories, cluttered stones.

I have only mind. Or only body.
Or none without the other. That is,
mind inhabits body which walks itself

of life. Like this dryas pasture going
to seed, white petals spinning into hair:
I spin like this, go on like this.

Sally Carricaburu

Curry Ridge

> *Is death*
> *by bear to be preferred*
> *to death by bomb?*
> —Maxine Kumin

Hiking to Curry Ridge, we burst out of woods onto tundra. Before us, a grunting bear cub. Adrenaline explodes like electric shock. I freeze, except my eyes, which frantically search the tundra and rocks. My thoughts cry *Where is she, where is she, the sow?* I imagine her tanklike body, fleet as a horse, her eyes, fierce as fire, flying at me, inexorably. My right hand clenches my bear spray, my thumb on its trigger. The cub does not see us. He toddles off, innocent as snow, as we slide behind a perimeter of rocks.

That night, I dream the sow, her fangs slicing muscle, cracking bone.

On Curry Ridge, we look across Ruth Glacier, eat jerky and pilot bread, drink Tang. In the hot sun, we follow cairns, stepping over fresh bear scat, to a fireweed-cloaked slope. I see the bears first. I spot them as we begin our traverse: one bear, then two, another sow with a cub, no—two cubs, another big one, blue black. I count eight bears. They graze on berries and move like waves across a pink-and-green sea. We slip along the slope beneath them, and though I fear them, they are lovely. In the eternal moment, we all waltz together like old friends.

Ann Dixon

Tree Bonking

When stout birches bow
beneath loads of snow
like supplicants, by grief
bent low, I cannot help

but sympathize.
One inch of snow, a sudden
freeze and some will snap.
Become dead trees.

I cannot halt
impending war, save
savaged children, one
oil-slicked shore.
A humbler task is set

for me. And so, alone
I shake and poke
till snow descends in heaps
of hope, releasing branches
that reach and rise, eagerly
toward bluest sky.

Marjorie Kowalski Cole

In the Shelter of the Forest

What shall we do for timber?
The last of the woods is down . . .
—FROM "KILCASH," ANONYMOUS, C. 1720,
TRANSLATED FROM THE IRISH BY FRANK O'CONNOR

SEVERAL YEARS AGO I dropped by to visit my mother in the house she bought in 1968, five miles southwest of Fairbanks. The house sits on the bank of the Chena River, which winds through Fairbanks on its way to the Tanana, which in turn empties into the Yukon.

In its lower stretches, the Chena is a cold, brown slough, home to beavers, ducks, kingfishers, terns, grayling, burbot, and king salmon. I sat down at her kitchen table, turned my head as always to take in the river, and shouted as anyone would have: "What happened to the trees?"

Across the water, several acres of spruce and birch forest had disappeared. The land had been cleared with heavy machinery. Chopped and splintered wood littered the ground. I looked over the brown wasteland at traffic roaring across the George Parks Highway bridge. The bridge had been there for years, behind the woodsy buffer; suddenly it was in her living room.

The sudden absence of trees is a shock. Her privacy was gone. This exposure not only dropped her home's value but also confused and hurt us. Having no choice, we put our minds to work on getting used to a revised landscape. Too soon, one struggles to remember what the earlier, tranquil world was like.

I began to understand, that day, how casually we treat the bounty and relative peace in which we're so lucky to live. But our view of things is changing rapidly. The confidence that once filled Alaskans, that our wilderness could not run out, has begun to evaporate. The cedars of Lebanon, the oaks of Ireland, the great pines of Michigan were also once considered endless yet have disappeared, just as many wildlife species in the continental U.S. have disappeared or been reduced to living in isolated "islands" of wilderness habitat separated by stretches of developed land.

From a plane over interior Alaska, we can still exult in unbroken vistas of wilderness. Mountain ranges are hundreds of miles away to the north and south. The interior itself is an expanse of taiga, wetlands, rivers, and rocky domes, with historic trails hidden below the trees. Dreaming about Alaska from afar, people compress its geography, as they do with any unknown place. But in fact, what is grand about the boreal forest, my home, is simply its spaciousness. It has largely been left alone.

But its character is changing. Contributing to the breakdown of this habitat is the clearing of forest for new construction and industry, along with the chaos associated with global warming: drought, fire, pests, and melting permafrost. Over a few generations we could lose this forest, even while forcing ourselves to adapt to incremental losses as each year goes by. Its wilderness qualities could move from daily experience into story, songbook, and ancestral memory—exactly what has happened in other parts of the world.

Ireland is such a place. My family has visited Ireland many times, finding it so pleasant that I was astonished to read, in older poetry, that it boasted a much different landscape not that long ago. It is largely treeless today, but ancient oak recovered from the bogs is evidence of a thriving, wood-based culture that once flourished there. Hardwood forests covered the island into Elizabethan times, and were lamented when they disappeared. "It was said that one could cross Ireland without touching the ground, so thick were the forests," said Irish writer Breandán Ó hEithir, and poet Seamus Heaney's mythic hero Sweeney almost does just that. But Irish oaks were cut down for Elizabethan warships, and the forests were cleared in order to eliminate hiding places for rebels, not unlike what happened with Agent Orange in Vietnam.

Since the eighteenth century, Ireland has been without forest cover despite a few parks, lone trees marking the edge of pastures, and Sitka spruce plantations maintained by the Board of Forestry. Like any monoculture crop, this imported species does its one job well. Sitka spruce supplies the building trade, but these trees do not create a forest. Instead, a plantation sticks out from the sides of a famous mountain like Ben Bulben like a bizarre haircut. Harvested, the wood lacks the strength of ancient trees. Warped two-by-fours at lumberyards are typical products of plantation growth.

Soil has eroded to rock in many parts of Ireland. County Galway's famous rocky Burren is actually, says poet and naturalist Moya Cannon, "an ecological disaster." Reminders of Ireland's native forests, as in the eighteenth-century poem "Kilcash," are a cause for wonder. This lyrical work by an anonymous Irish-speaking writer laments the diversity and abundance that is lost when a forest disappears—the insect and bird life, the game, the varied species, the music of a living forest:

> There's mist there tumbling from branches,
> Unstirred by night and by day,
> And darkness falling from heaven,
> For our fortune has ebbed away,
> There's no holly nor hazel nor ash there,
> The pasture's rock and stone,
> The crown of the forest has withered,
> And the last of the game has gone.

With the forest, an entire community vanished. And within a few generations, Alaskans too could be living in a drastically revised landscape.

Conrad Richter's historical novel *The Trees* revisits pioneer days in the Ohio River Valley. Passages from the novel are startling to modern readers who have never experienced a great forest firsthand. Richter's heroine, Sayward, gets her first look at Ohio territory from a ridge:

> For a moment Sayward reckoned that her father had fetched them unbeknownst to the western ocean and what lay beneath was the late sun glittering on green-black water. Then she saw that what they looked down on was a dark, illimitable expanse of wilderness. It was a sea of solid treetops broken only by some gash where deep beneath the foliage an unknown stream made its way. As far as the eye could reach, this lonely forest sea rolled on and on till its faint blue billows broke against an incredibly distant horizon. . . .

In Alaska, thousands of years of localized use, including the past century of rapid development, has not yet destroyed the forest cover. As a panorama this landscape is never a monotone sea of treetops, such as Sayward observed, but is always broken into the mosaic of forest types and colors so characteristic of taiga around the globe; every few miles the forest is reacting to slightly different amounts of water, nutrition, and solar radiation, or is in another stage of recovery from fire or flood.

Interior Alaska lies within the circumpolar, subarctic boreal forest that stretches through Siberia, Canada, and Scandinavia. The Taiga Rescue Network, based in Sweden, estimates that five to six percent of America's old-growth forest is intact today in the continental United States—which is another way of saying we have removed ninety-five percent of our native forest cover. But when Alaska is included in the statistics, Americans still have about fifteen percent of their native uncut forest, thanks largely to this swath of taiga.

A day tour around Fairbanks leads from stands of paper birch mixed with white spruce, to platoons of twisted black spruce and tamarack in poorly drained flats over discontinuous permafrost, to slopes where skinny balsam poplar and quaking aspen rise almost like weeds from thin soil. The climate and soil conditions support a short list of conifers and hardwoods. Only six species of trees are native to Alaska's interior, compared to the thousands you might find in a few square miles of Amazonian jungle.

Growing in the best-drained, sunniest locations, white spruce are the tallest, straightest, and clearest of Alaska's timber trees. Slower growing than hardwoods, white spruce often begin their lives under a canopy of birch leaves. On a June morning the dark spruce and newly leafed-out birch present a lovely and characteristic sight in the hills.

Black spruce struggle for existence on north-facing slopes or in icy lowlands. These trees often have a stunted look. Each one develops a unique shape as it accommodates to its own few square feet of nutrient-poor soil. They expand their population by layering: the lowermost branches, touching the ground, form roots and become new upright stems, a phenomenon that can make a black spruce forest almost impossible to stroll through.

Lone tamaracks, belonging to the larch family, are found here and there among the black spruce. An evergreen that loses its needles in the fall, a tamarack in winter looks like an abandoned hat rack in the woods. Its brown twigs with clusters of berrylike cones add a spectral note to a stand of black spruce.

Often the cold lowlands of the interior become a weirdly beautiful scene. Ghostly yet full of life, they are habitat for beavers, muskrats, bear, moose, and dozens of species of birds.

Boreal trees have a fairly short time each year in which to use the soil and to contribute their own organic matter to it. In recent years, foresters have studied how fungi play a vital part in forest health within the soil. Parasitic fungal strands called mycorrhizae absorb nutrients and water from the soil and pass them directly to the tree root. In return, the host tree provides sugars from its own photosynthesis to the fungi. Mycorrhizae were discovered about a hundred years ago, but have only been studied intensively in the last thirty years as part of an interest in the possibilities of reforestation.

Animals of the boreal forest display an amazing variety of adaptations to the climate. Black-capped chickadees are so polite at backyard feeders (giving way to rude, assertive redpolls) that one wonders if the chickadees are getting their share. But in their exceptional memories they store the locations of seed caches throughout their territory. After visiting each cache during daylight hours, they huddle down for sleep, burn every calorie, and start in again the next day.

The silence of the boreal forest in winter is more like the quiet intervals within a symphony than a deathlike stillness. Approximately thirty species of birds overwinter in interior Alaska, including owls, hawks, ravens, jays, woodpeckers, and songbirds. The list grows longer as individuals from warmer latitudes find their way north and linger into December. Robins, magpies, and ducks are now surprising sights in the winter.

How the boreal ecosystem works is not fully known to forest scientists. All of its secrets are not translated from the indigenous languages of those who have lived here for thousands of years, the Native people and subsistence users of the far north. We do know, however, something of the importance of the forest to the health of our atmosphere. The circumpolar boreal forest, including that sizable chunk within the United States, is no less important for the exchange of carbon dioxide and oxygen on this planet than the tropical rain forests. And we know that this healthy band of forest around the globe is threatened by large-scale logging and by the effects of global warming.

This boreal forest has always functioned through cycles of growth, decay, collapse, and rebirth, set off by local disturbances such as lightning-caused forest fires, floods, and insect infestations. These events are critical to the system. They give new growth an opportunity and contribute nutrition to the soil. Lightning-caused fire is the primary catalyst, sometimes burning several million acres in Alaska every summer, and regrowth from fire has resulted in a mosaic of specific forest types throughout the taiga. But fires, insects, and melting permafrost have reached new extremes, going beyond the boreal system's capacity to rejuvenate itself.

In Canada, Siberia, and Scandinavia, industrial-scale logging has also damaged forest cover. Alaska's mills have not reached that scale, but the threat continues to surface. Years ago, lumbermen might have laughed to think of logging these latitudes for anything but fuelwood and local housing needs. There are no behemoths north of the Alaska Range; the tallest white spruce rarely top one hundred feet. But as demand for wood pulp grows worldwide, the boreal forest has come under severe pressure. Twenty-first-century pulp mills have tremendous appetites. Sections of forest in other northern countries are being chewed and digested to make tissues, paper towels, and other low-grade products in today's system of exploitation by large multinational corporations offering local wages in exchange for a steady supply of raw material.

At present, the most attractive tree to the logging industry is the white spruce. Logging occurs on a fairly moderate scale here, but more and more frequently trucks from the southern parts of the Tanana River Valley barrel into Fairbanks with white spruce bound for export. Each log was once a climax tree—a condominium of bird, insect, and small mammal habitation, an inhabited skyscraper holding a piece of the wild.

In 1994 a state senator from Fairbanks proposed large-scale logging concessions in Alaska's interior, with a bill that tried to fling wide the door to industry. Residents overwhelmingly opposed this action, with hearings and testimony from across the political spectrum that took the legislature by surprise. After several months the bill died and the months of activism were a lesson not lost on either side of the issue. Foresters writing about timber harvest over ten years later found the uproar worthy of serious consideration. An essay in the recent book *Alaska's Changing Boreal Forest* notes that "Even a modest proposal for timber harvest is the subject of intense attention by the local public. . . . The state has occasionally deferred timber sales in response to public opposition."

One way to avoid public outcry is to keep the public little-informed, to involve the public as little as possible. Such detachment threatens to leave local citizens at the mercy of forces beyond our influence. Everywhere this detachment occurs—from Nigeria with its oil development, to southern Louisiana with its loss of coastal wetlands—environmental protection clearly has become a human rights issue.

Years after the initial outrage, legislators quietly readjusted a couple of outcomes. In 2003, they added "pulping of timber" to the list of high value-added timber uses, putting tissue paper right up there with local home building and furniture making. At the same time, they rewrote the forest management plan for the Tanana Valley State Forest to make logging its primary purpose, thus throwing out the "multiple use management" overwhelmingly favored by citizens.

Global warming is recognized as a primary and overarching danger to the environment, particularly in Alaska. It not only leads to deforestation worldwide but is dramatically worsened by it. Tree cover is a moderating influence on the climate; as trees disappear, the problem becomes more severe. Statistics, anecdotes, models, scientific analyses, and linguistic evidence that global warming is hurting Alaska's forests pile up year by year. Damp lowlands expand as permafrost melts and drowns low black spruce forest. Ducks huddle on unfrozen stretches of the Chena River all winter long. Arid summers slow white spruce growth. Beetles and other insect pests wreak havoc. Up and down the Yukon River, plugs drilled from living spruce show the extent of the damage.

The good news is that we seem to be getting the message that the earth is indeed one system, not truly divisible by national boundaries. The health and stability of the atmosphere is perhaps the most important reason that our forest should be left whole, logged selectively for regional needs with regrowth always part of the equation. Human nature resists the lessons of history, but we are learning to see things globally.

Forests are the lungs of the earth. The role of the boreal forest in the global exchange of carbon and oxygen is reason enough to give this planet a Sunday off, a moratorium, so to speak, on industrial logging. Alaskans are used to unbroken vistas of wilderness. From an airplane, there's so much of it. The forest is endless, the lights of cabins and villages still few. How can we run out?

We need to make forest size and forest health our top priorities, to see this ecosystem not so much in agricultural terms, as yielding a useful crop, but rather as part of this planet's air-conditioning system, an assist to the atmosphere itself in keeping us alive and healthy. And a healthy, sizable forest feeds our imagination and our spirits: this is clear from literature and religion and the cultures of all once-forested places. Most of all, a forest helps to maintain life.

Trees are amazing systems in and of themselves. They transmit water from base to leaf-tip against the force of gravity and they make use of all available sunlight. In pampered conditions, individual white spruce and birch attain height and diameter unknown in the midst of the forest. Landowners delight in favorite trees, in wide birch and tall straight spruce. Handsome individuals decorate the streets of my hometown. A fat birch tree sets out a canopy of tiny, heart-shaped leaves over a hammock in the spring, and lays down a golden carpet from front door to mailbox in the fall. In October, fallen leaves fade to a pinkish brown and, if not covered by snow, lend a rosy alpenglow to the long subarctic twilights. White spruce grow tall and symmetrical in front yards, ready to take on a hundred or more feet of Christmas lights.

But trees don't exist to grow singly. As novelist and naturalist John Fowles wrote:

> Trees are no more natural as isolated specimens than man is as a marooned sailor or hermit. Their society in turn creates or supports other societies of plants, insects, birds, mammals, micro-organisms; all of which we may choose to isolate and section off, but which remain no less the ideal entity, or whole experience, of the wood. . . .

The health of individual trees in suburban settings, in short, should not necessarily assure us that the forest itself can recover from industrial clearcuts. Trees grow back after a disturbance, but there is a difference between a local, natural disturbance, and a massive, industrial clearcut. The first event leaves nutrients on the ground, and the forest accommodates. The heat of fires, for instance, opens seed packages of black spruce. The second, a clearcut, scrapes away wheat and chaff alike. The incredible efficiency of modern logging operations, the world's appetite for disposables and convenience, and the sense local people have that steady employment is only to be found with major corporations make a dangerous combination.

The same state senator who proposed the logging concessions in 1994 bought the cleared land across the Chena River from my mother's house and built a tourist camp on the spot, with hundreds of little cabins and RV hookups. He planted grass on the denuded riverbank, much of which slipped into the river within a couple of months. The cabins are luxurious, although jammed together, so that the whole place resembles one of the enlightened migrant labor camps described in John Steinbeck's *Grapes of Wrath*. Road-weary tourists who bed down for a few days in these cabins will not miss what they've never seen—the forest that sheltered local people and wildlife for generations.

They will not miss the countless gifts of the forest that come to individuals fortunate enough to live near or within its borders—its shade and peace, wildlife, mushrooms, berries, flowers, and herbs, its mystery and profound lessons. They will not miss those moments that have inspired Alaska's artists and poets, such as John Haines, whose 1966 collection *Winter News* proved once again that to listen, absorb, honor, and note every detail, in short to take an inventory of forest life rather than to conquer it, could lead to beautiful poetry and increase our understanding of the environment.

Forests have always been places to hide, but more than that, they loan their riches and their mystery to every human community nearby. It is not certain that we can survive without the shade and shadow of the great forests, without their silent labor on our behalf. One would love to say to all the logging interests, pulp mills, veneer plants, and multinationals that come tapping at our doors: Go away, we're closed for inventory. We need this place.

Over my desk is posted a scrap of poetry from an anonymous Irish scribe in the ninth century, a monk writing in a landscape now vanished:

> A wall of forest looms above
> And sweetly the blackbird sings.
> All the birds make melody
> Over me and my books and things . . .

The boreal forest is a living organism that offers us a chance to admit to our sense of wonder and gratitude. Its trails tempt us to enter farther in. May they also tempt our grandchildren.

Mike Burwell

Wash Silver

When I slept in the woods I woke before dawn and drank brandy
And listened to the birds until the moon disappeared.
—Jim Harrison

I'm lying if I say the moon ran me, if it did more than spend
its pale time in the sky, hung, or the birds—ravens' clickings—
did more than hang briefly in my ears. And there was no brandy,
only beers smuggled in on planes from Ketchikan.
I did not run with beauty but cut a road's bright claw through spruce
and rain, watched the trees fall behind me, the earth dug and pushed
away by road builders' wet, green hunger for the island.

But there were evenings in August when sun returned, and I drove
to the dark cuts in the high blueberries along the road, followed
these paths to the river that flowed from the deep life of Klawock
Lake. I threw in the orange and pink day-glo spinners that I,
needing no luck, used to hook the wild cohos that hit and fought
and ran, most leaping back into the silver mystery of their food
because I would not buy a net to land their dancing.

Some nights with a hooked fish slapping downriver
and the brush trilling with birds, I'd see the whole scene wash silver,
see these slicing fish hurtle like errant moons through the late,
deep light.
And I drank the thick bright milk of it, all of it.

Richard Nelson

The Forest of Eyes

AFTER A LONG HIKE, taking the easy routes of deer trails, my dog, Shungak, and I move into a stand of shore pine that ends beside a half-overgrown logging road. This is the first sign of human activity since we left camp, and it indicates we're approaching the clearcut valley. The road follows a narrow band of muskeg that has all the delicate loveliness of a Japanese garden, with reflecting ponds and twisted pines in bonsai shapes. Farther on, it cuts through an alder thicket and runs up a steep, forested slope. A dense flock of birds sprays into the high trees, twittering like canaries, hundreds of them, agitated and nervous, moving so quickly they're difficult to hold for long in the binoculars.

The birds are everywhere, hanging upside down from the twigs and working furiously on spruce cones. Each one plucks and twists at its cone, shaking loose the thin scales and letting them fall. The air is filled with a flutter of brown scales. I recognize the sparrow-sized pine siskins immediately, then identify the larger birds as white-winged crossbills. I've never had a good look at a crossbill before, but the hillside roadway gives an easy view into the tree crowns, where bright red males and olive females swarm through the boughs. With some patience, I can discern the tips of their beaks, which crisscross instead of fitting together like an ordinary bird's. This allows the crossbill to pry the scales apart and insert its tongue to extract the seeds embedded deep within. Once again, I'm reminded that tropical animals aren't the only ones who have added a little adventure to their evolution. Suddenly the whole flock spills out from the trees and disappears, like bees following their queen. After another half mile, a slot appears in the road ahead. As we approach, it widens to a gateway out of the forest—a sudden, shorn edge where the trees and moss end, and where the dark, dour sky slumps down against a barren hillside strewn with slash and decay.

Oversized snowflakes blotch against my face and neck, and the breeze chills through me. I look ahead, then look back toward the trees, breathless and anxious, almost wishing I hadn't come. It's the same foreboding I sometimes feel in the depths of sleep, when a blissful dream slowly degenerates into a nightmare; I am carried helplessly along, dimly hoping it's only a dream, but unable to awaken myself and escape.

The road angles into a wasteland of hoary trunks and twisted wooden shards, pitched together in convulsed disarray, with knots of shoulder-high brush pressing in along both sides. Fans of mud and ash splay across the roadway beneath rilled cutbanks. In one place, the lower side has slumped away and left ten feet of culvert hanging in midair, spewing brown water over the naked bank and into a runnel thirty feet below.

A tall snag clawed with dead branches stands atop the hill. I decide to hike up toward it rather than walk farther along the road. At first, it's a relief to be in the brush, where I can touch something alive, and where my attention is focused on the next footstep rather than the surrounding view. But thirty yards into it, I realize that moving through a clearcut is unlike anything I've ever tried before. The ground is covered with a nearly impenetrable confusion of branches, roots, sticks, limbs, stumps, blocks, poles, and trunks, in every possible size, all gray and fibrous and rotting, thrown together in a chaotic mass and interwoven with a tangle of brittle bushes.

An astonishing amount of wood was left here to decay, including whole trees, hundreds of them in this one clearcut alone. Some flaw must have made them unusable even for pulp, but they were felled nonetheless, apparently so the others would be easier to drag out. Not a single living tree above sapling size stands in the thirty or forty acres around me.

I creep over the slippery trunks and crawl beneath them, slip and stumble across gridworks of slash, and worm through close-growing salmonberry, menziesia, and huckleberry. Even Shungnak struggles with her footing, but she gets around far better than I do, moving like a weasel through a maze of small holes and tunnels. I can tell where she is by the noise she makes in the brush, but only see her when she comes to my whistle. In some places I walk along huge, bridging trunks, but they're slick and perilous, and I risk falling onto a deadly skewer of wood below. I save myself from one misstep by grabbing the nearest branch, which turns out to be devil's club, festooned with spines that would do credit to any cactus. We also cross dozens of little washes that run over beds of coarse ash and gravel. There are no mossy banks, no spongy seeps, just water on bare earth. By the time we near the top I am strained, sweating, sore, frustrated, and exhausted. It has taken me almost an hour to cross a few hundred yards of this crippled land.

I've heard no sound except my own unhappy voice since we entered the clearcut, but now a winter wren's song pours up from a nearby patch of young alders. I usually love to hear wrens, especially during the silence of winter. But in this topsy-turvy place the reedy, contorted phrases, rattling against the beaten hill, seem like angry words in some bewildering foreign tongue. I picture a small, brown-skinned man, shaking his fist at the sky from the edge of a bombed and cratered field.

A large stump raised six feet above the ground on buttressed roots offers a good lookout. The man who felled this tree cut two deep notches in its base, which I use to clamber on top. It's about five feet in diameter and nearly flat, except for a straight ridge across the center where the cutter left hinge wood to direct the tree's fall. The surface is soggy and checked, but still ridged with concentric growth rings. On hands and knees, nose almost touching the wood, using my knife blade as a pointer, I start to count. In a short while, I know the tree died in its four hundred and twenty-third year.

I stand to see the whole forest of stumps. It looks like an enormous graveyard, covered with weathered markers made from the remains of its own dead. Along the slope nearby is a straight line of four stumps lifted on convoluted roots, like severed hands still clasping a nearly vanished mother log. Many of the surrounding stumps are smaller than my platform, but others are as large or larger. A gathering of ancients once stood here. Now it reminds me of a prairie in the last century, strewn with the bleached bones of buffalo. Crowded around the clearcut's edges are tall trees that seem to press forward like curious, bewildered gawkers.

Two centuries ago, it would have taken the Native people who lived here several days to fell a tree like this one, and weeks or months to wedge it into planks. Earlier in this century, the handloggers could pull their huge crosscut saws through it in a couple of hours. But like the Native Americans before, they selected only the best trees and left the others. Now I gaze into a small valley miles deep, laid bare to its high slopes, with only patches of living timber left between the clearcut swaths.

Where I stand now, a great tree once grew. The circles that mark the centuries of its life surround me, and I dream back through them. It's difficult to imagine the beginnings—perhaps a seed that fell from a flurry of crossbills like those I saw a while ago. More difficult still is the incomprehensible distance of time this tree crossed, as it grew from a limber switch on a forest floor to a tree perhaps one hundred and fifty feet tall and weighing dozens of tons. Another way to measure the scope of its life is in terms of storms. Each year scores of them swept down this valley—thousands of boiling gales and blizzards in the tree's lifetime—and it withstood them all.

The man who walked up to fell this tree some twenty years ago would have seemed no more significant than a puff of air on a summer afternoon. Perhaps thin shafts of light shone down onto the forest floor that day, and danced on the velvet moss. I wonder what that man might have thought, as he looked into the tree's heights and prepared to bring it down. Perhaps he thought only about the job at hand, or his aching back, or how long it was until lunch. I would like to believe he gave some consideration to the tree itself, to its death and his responsibilities toward it, as he pulled the cord that set his chainsaw blaring.

The great, severed tree cut an arc across the sky and thundered down through its neighbors, sending a quake deep into the earth and a roar up against the valley walls. And while the tree was limbed and bucked, dozens of other men worked along the clearcut's advancing front, as a steady stream of trucks hauled the logs away.

A Koyukon man named Joe Stevens once took me with him to cut birch for a dogsled and snowshoes. Each time we found a tall, straight tree with clear bark, he made a vertical slice in the trunk and pulled out a thin strip of wood to check the straightness of the grain. When we finally came across a tree he wanted to cut, Joe said, "I don't care how smart a guy is, or how much he knows about birch. If he acts the wrong way—he treats his birch like it's nothing—after that he can walk right by a good tree and wouldn't see it." Later on, he showed me several giant, old birches with narrow scars on their trunks, where someone had checked the grain many years ago. In the same stand, he pointed out a stump that had been felled with an ax, and explained that Chief Abraham used to get birch here before the river made a new channel and left his fish camp on a dry slough.

Joe and I bucked the tree into logs and loaded them onto a sled, then hauled them to the village and took them inside his house. It was important to peel the bark in a warm place, he said, because the tree still had life and an awareness in it. Stripping the log outside would expose its nakedness to the winter cold and offend its spirit. The next day, he took the logs out and buried them under the snow, where they would be sheltered until he could split them into lumber. Later on, when Joe carved pieces of the birch to make snowshoe frames, I tried to help him by putting the shavings in a fire. His urgent voice stopped me: "Oldtimers say we shouldn't burn snowshoe shavings. We put those back in the woods, away from the trails, where nobody will bother them. If we do that, we'll be able to find good birch again next time."

The clearcut valley rumbled like an industrial city through a full decade of summers, as the island's living flesh was stripped away. Tugs pulled great rafts of logs from Deadfall Bay through tide-slick channels toward the mill, where they were ground into pulp and slurried aboard ships bound for Japan. Within a few months, the tree that took four centuries to grow was transformed into newspapers, read by commuters on afternoon trains, and then tossed away.

I think of the men who worked here, walking down this hill at the day's end, heading home to their families in the camp beside Deadfall Bay. I could judge them harshly indeed, and think of myself closer to the image of Joe Stevens, but that would be a mistake. The loggers were people just like me, not henchmen soldiers in a rebel army, their pockets filled with human souvenirs. They probably loved working in the woods and found their greatest pleasure in the outdoors. I once had a neighbor who was a logger all his life, worked in these very clearcuts, and lost most of his hearing to the chainsaw's roar. He was as fine a man as I could hope to meet. And he lived by the conscience of Western culture—that the forest is here for taking, in whatever way humanity sees fit.

The decaying stump is now a witness stand, where I pass judgment on myself. I hold few convictions so deeply as my belief that a profound transgression was committed here, by devastating an entire forest rather than taking from it selectively and in moderation. Yet whatever judgment I might make against those who cut it down I must also make against myself. I belong to the same nation, speak the same language, vote in the same elections, share many of the same values, avail myself of the same technology, and owe much of my existence to the same vast system of global exchange. There is no refuge in blaming only the loggers or their industry or the government that consigned this forest to them. The entire society—one in which I take active membership—holds responsibility for laying this valley bare.

The most I can do is strive toward a different kind of conscience, listen to an older and more tested wisdom, participate minimally in a system that debases its own sustaining environment, work toward a different future, and hope that someday all will be pardoned.

A familiar voice speaks agreement. I squint up into the sleet as a black specter turns and soars above, head cocked to examine me. A crack of light shows through his opened beak; his throat fluffs out with each croak; downy feathers on his back lift in the wind; an ominous hiss arises from his indigo wings. Grandfather Raven surveys what remains of his creation, and I am the last human alive. I half expect him to spiral down, land beside me, and proclaim my fate. But he drifts away and disappears beyond the mountainside, still only keeping watch, patient, waiting.

I try to take encouragement from the ten-foot hemlock and spruce saplings scattered across the hillside. Interestingly, no tender young have taken root atop the flat stumps and mossless trunks. Some of the fast-growing alders are twenty feet tall, but in winter they add to the feeling of barrenness and death. Their thin, crooked branches scratch against the darkened clouds and rattle in the wind. The whole landscape is like a cooling corpse, with new life struggling up between its fingers. If I live a long time, I might see this hillside covered with the beginnings of a new forest. Left alone for a few centuries, the trees would form a high canopy with scattered openings. Protected from the deep snows of open country, deer would again survive the pinch of winter by retreating into the forest. The whole community of dispossessed animals would return: red squirrel, marten, great horned owl, hairy woodpecker, golden-crowned kinglet, pine siskin, blue grouse, and the seed-shedding crossbills. In streams cleared of sediment by moss-filtered runoff, swarms of salmon would spawn once more, hunted by brown bears who emerged from the cool woods.

There is comfort in knowing another giant tree could replace the one that stood here, even though it would take centuries of unfettered growth. I wish I could sink down into the earth and wait, listen for the bird voices to awaken me, rise from beneath the moss, and find myself sheltered by resplendent boughs. And in this world beyond imagination, such inordinate excess toward nature will have become unthinkable.

Anne Coray

Sweet Drug of the Backward Drag

Let the moon be my elixir,
my liniment the calm.
Let the mountains lit mauve at dusk
act as sedative to spirit, the sighting of swans

gather like powder in dreams.
And the lynx that crept down the path
in the pool of a drowsy evening
ease the muscle and taut skull.

How far we have come from the soul's root,
the rare Earth's nebula,
pollen and white seed. How far
will we go with the vice of our inventions?

Whose voices, borne by wire
from dammed-up waters, whose fingers, transistor-
switched, whose hearts' arrhythmias turned rhythm,
charged with a fevered blood?

Evolution is a one-way street,

amphetamine from which we never branch or furl;
we've lost the simple map, whose X marked here,
meaning back, the route that meandered downriver
through cottonwood and fern,

while the last leaves' spin in autumn
quieted the nerves and reset the clock
to the hands of the four directions; the leaves
pausing in their descent,

in their descent, pausing
to point east, north, south, west
before settling, bright stars, on a current
bound for the motherland, the ancient sea.

Susan Pope

Losing Out to Baseball, Motherhood, and Apple Pie

WE NEIGHBORS LAID THE TRAIL OURSELVES, not with picks or shovels, but with meandering feet, seeking solitude in our backdoor piece of nature. We believed the park belonged to all of us, just like it was.

We knew our fellow trailblazers mostly by their tracks. An eager pair of skiers broke trail in the watery light of a January dawn. After the big Chinook, a brave snowshoer scraped atop the icy snow crust, occasionally breaking through to the thawing brown muck below. Two families hauled sleds full of children across a fresh Easter morning snowfall while their dogs plowed wide arcs off the trail, sniffing out voles buried deep below.

Over many years, we carved a narrow furrow through the black spruce swamp, working our way up through the rolling hills of paper birch, and down through the slowly filling kettle pond. The straight young alders provided easy handholds to haul ourselves up the ridge. There we crawled over the windfalls of beetle-killed spruce, and slowly circled back to the street to rejoin the frenzy of daily life. Occasionally, our dogs tangled with each other—sniffing, snarling, scrapping, until one proved tougher or a bigger bully, or we pulled them apart, lying to each other, "He's usually not like this."

My husband, Jim, and I cast our stories out to the trees and bushes along the little trail. Step by step, our bodies loosened and our voices warmed. We raised a daughter, mourned dead parents, patched up a marriage more than once, fought each other, fought back the world, sought peace, and rejoiced in the birth of our grandchildren and the return of the ruby-crowned kinglet each spring. Who knows what damage Jim and I would have done to each other or the world without our spring slogs through the swamp or our crunchy snowshoe plods through fresh snow. We cursed, laughed, cried, fell in love, and prayed with wandering feet and open eyes.

We were naive. We saw rolling hills, trees, and birds' nests. Other people, though, saw the park differently—a blank slate, a place of "potential." They saw an open field with baseball diamonds, cheering crowds, bleachers, a picnic shelter, snack bar, playground, and a parking lot. Two views as far apart as July and January.

We neighbors and our beehive of well-organized Friends of the Park believed that since we had the law, a good attorney, righteousness, and the Internet working for us we would prevail. Leaflets posted on car windshields all over town, proclaimed

> We're not against
> Baseball, Motherhood, and Apple Pie!
> We just want to save our park!

We rallied, wrote letters to the editor, held meetings, outnumbered our opponents at every opportunity for public testimony, but in the end we underestimated the power of political deals and sad little boys and girls in baseball jackets.

In 1974, roughly four thousand acres of federal land were turned over to the city of Anchorage to create a park commemorating the nation's 1976 bicentennial. Once owned by the Department of Defense, the land was crisscrossed by overgrown jeep trails and littered with pockets of rusting C-ration tins and empty five-gallon gas cans. By then, hikers, berry pickers, skiers, and dog mushers had begun to share the land with the indigenous moose, bear, coyotes, lynx, and other wildlife. The old Borough Assembly (now called the Anchorage Assembly, representing a unified city and borough government) called on a smattering of "outdoor types" to help them craft a park plan. Since it was to be Anchorage's piece of our nation's two-hundredth birthday, it was named Far North Bicentennial Park. The land had long been coveted by Anchorage residents as the final link in a wild corridor joining the Chugach

Mountains in the east to Cook Inlet in the west, along the muddy banks of Campbell Creek. The planners, peering over the brink of the oil explosion about to hit the state, sought to preserve a piece of open space in a city confined to a glacial bowl between the mountains and the sea. Since then, the population of Anchorage has steadily swelled, devouring dry and wet land, creeping right up to the rim of the park. Pressure on the city to give up pieces of the land crescendoed with every building boom. By the time the park plan came up for its first renewal in 1980, a loose collection of users joined together to prevent a piecemeal dismantling of the woodland. They called themselves the Friends of Far North Bicentennial Park.

Ours is an old neighborhood by Anchorage standards. The city offered lots in our subdivision to Anchorage families who had lost their homes, land, and even their loved ones in the 1964 earthquake. Each house sprouted separately over the years, planted on large lots where builders carefully preserved clumps of trees from the old forest. When we moved here twenty years ago, the forest encircled three sides of our cluster of some eighty homes that were perched on the western boundary of Far North Park. One summer when my mother's family was visiting from New York, I drove my Uncle Frank on a tour of the neighborhood. He asked, "Why are all the houses different?" I had never before noticed the range of our houses: A-frames, compact one-story boxes, 1970s split-levels, sprawling cedar-sided ranches with attached auto shops.

A corps of old-timers banded together every few years to ward off additional stripping of the old forest. Our victories were small—a tight new batch of homes instead of a sprawling church and parking lot, a thin line of trees to buffer us from the new neighbors, a dead-end street instead of an open artery pumping traffic into our unpaved streets. Gradually, we became a ragged island in the suburbs, encircled by a sleek new school, a million-gallon water tank "camouflaged" with painted spruce trees, and yet another upscale development of massive earth-tone homes plopped on dinky squares of denuded land. As earthmovers scratched away the forest and wiped out our well-trod paths, more and more neighbors sought silence in the back trail through the swamp. We still had the park.

"Times change," my daughter Elisha said during one of my rants. "Get a grip, Mom."

She was right. We would lose trees, not children, or grandchildren. But a forest is more than a collection of trees. I would lose a piece of my family.

"Besides," she pointed out, "it's only twenty-five acres, and those kids need a place to play ball."

"But why do they want to take away *my* place so that they can have *their* place?" I argued.

She rolled her eyes.

She played soccer as a kid, before she went out for boys. Her team played on high school fields. I never considered that her team deserved its own field. So I was shocked by the fervor of the Little League parents in pursuit of the "right" to their own fields. For them the cause was simple: kids versus greenies. We were elitist tree huggers who wanted to lock up the land for ourselves. They were families who just wanted a place for their kids to play ball.

Even though Friends of the Park outnumbered the Little Leaguers and their parents at every public hearing, the Little Leaguers' presence could not be ignored—bright nylon team jackets and baseball caps, bored kids scuffing and coughing through the impassioned pleas of their parents. They were about to become homeless—evicted from their fields by the rich developer who had allowed them to play on his land for decades. Yet, *he* was not the enemy. We were the kid haters denying them a new home.

"Wait," I wanted to yell during a long night of repetitive testimony. "I'm a mother and grandmother. How can you say I hate kids?" But I had already taken my turn to speak.

Throughout our bitter tug-of-war over the land, we Friends of the Park suspected the city administration had bigger plans for the four-thousand-acre park than just Little League fields. The stakes were high. Carving out a chunk for ball fields would mean challenging the intent of the ordinance that designated the land as a "natural park." It meant opening up the possibility for other development in the future. First ball fields, then roads, then houses, then strip malls.

The city decided to buy time, placate the Friends of the Park, and cool off the media heat by "studying" alternative sites. Since the park had survived a similar threat several years ago when a group sought to build a cultural center, we Friends thought that with organized opposition, reasoned arguments, and overwhelming show of force we could defeat the demand. Besides, there were plenty of school fields lying idle most of the summer. But, in the end, the long and expensive city-funded search for alternative sites meandered like our little trail and wound up back in the old neighborhood. The twenty-five-acre chunk of our park had been the Little Leaguers' all along.

Behind the scenes, a petition to put the issue on the spring election ballot popped up. The Friends started a counter lawsuit to keep it off the ballot. Meanwhile, Anchorage Assembly members started working on a deal to extinguish the political forest fire that threatened to burn up their meetings, email boxes, and political careers. Two of them proposed an ordinance that would allow the fields to be built, in return for a ban on future development. Finally, just before Christmas of 2002, the Friends mobilized their final protest, their last appeal to preserve the integrity of the park. More than forty people testified to the Anchorage Assembly; most were opposed to the ordinance.

"It's not a fair trade," the Friends argued. A precedent was being set, opening the door for a future group to plead their cause to the assembly and overturn the ban on development. The ordinance passed by a vote of 6–4.

The following summer, two women expressed their grief by tying cloth ribbons around the doomed trees. Word went out by email; people filled their pockets with strips of cloth and walked our old trail. Soon, pink, red, and yellow streamers rippled in the breeze. Someone passing by might have thought it was the path to a wedding or a midsummer festival. But attached to the trees were poems, pictures, farewells, and epitaphs such as "a lynx paused here last spring" and "here lies a downy woodpecker's nest." It was our last protest, a resigned goodbye to old friends. As I wove among the trees along our familiar path I felt survivors' guilt—depressed, disloyal, ashamed that I would survive, find another trail, go to work, eat, drink, sleep under a roof, and drive my car to the mall. I had not tried hard enough to save the patch of woods that had given me so much.

I couldn't bear to witness each tree being chewed up and coughed out in chips by the HydroAxe. During the weeks of rumbling, grinding, and high-pitched whining that reverberated through the neighborhood, I stayed away. One Sunday morning while the machines were resting, Jim and I followed the old trail. It seemed intact, and I tried to convince myself that nothing had changed. But right where we would have begun our climb up the ridge through the alders we stumbled into a nightmare. Our ridge had vanished, the forest ripped apart into an oozing muck of amputated limbs, jagged stumps, and twisted sticks. It was then that my denial switched to rage.

A few days later I cheered when I picked up the local paper and read how an elderly neighbor had been arrested for assault after jabbing a tree cutter with a cane. I too wanted to relieve my feelings of powerlessness.

Late last fall, curiosity carried me back to our spot. Viewing the corpse might release me, I reasoned. It didn't help. I found nothing. The terrain was scraped clean, leveled, barren—all that was familiar had been erased. A newcomer would drive by this naked land and never know that a forest once grew here, would never know its stories, would never know the trail that once rolled over the old contours of the land.

After that, I stayed away.

In November, eight inches of new powder carpeted a base of six inches of packed snow. The temperature hovered near twenty degrees. I convinced Jim that conditions were perfect; we had to ski. After dinner we dug out our gear from summer hiding places, zipped jackets, slipped on hats and mittens, and sealed up the Velcro fronts of our gaiters. We hoisted our skis and poles and plodded the half mile through our neighborhood's dark, unplowed streets, then passed beneath the bright, steady streetlights of the new neighborhood, and plunged through the calf-deep snow that hid the bike trail to the school. There we snapped on our skis, slid our mittens through the hand loops on our poles, and began our glide toward Far North Bicentennial Park.

At first I labored to find a rhythm. My breath puffed out in short bursts of vapor while my heartbeat echoed in my ears. But soon my old fish-scale skis carried me forward in their own dance, alternately gripping and sliding in the perfect snow. Jim skied far enough ahead to give me solitude and a chance to cast off the worries of my workday. I herringboned to the top of the hill, unzipped my jacket, and plunged down a long, curving slope. I welcomed the cool wind on my chest.

Jim reached the clearing long before I did. I crested the top of the old hill where the alders once arched a shelter around our path. A blast of wind slapped my face. I shook away an icy tear zipping across my cheek.

Traffic racket rolled unfiltered across the stripped field. The city lights lay exposed below me. It was as if we had instantly joined the city. I guess we had.

I stopped to watch Jim's dark form skim across the blank snow. Next summer children will play ball here, I thought. Their parents will cheer them on. Families will eat hot dogs and drink lemonade and be thankful that at last they have a park of their own. They are not evil people. They will be content with the groomed grass and paved parking. They will not miss the little trail they never knew or the joy of quiet wandering, pausing to search for a varied thrush whistling somewhere in the black spruce. But I do. And I miss the comfort of that twisting path where I could lose myself and find my way back out again.

Ken Waldman

Garbage Bears, Hoonah

I heard how one enterprising tour guide
drives summer visitors to the dump
to see local grizzlies paw mounds of trash
as they feed on rinds, peels, seeds, bones.
And how the high schoolers approach
fearlessly in parents' pickups, honking horns,
lobbing beer cans, occasionally catching
a slow one from behind, front bumper first.
Because I collect stories, I asked
to be taken, and was, the night after
the monsoon, gravel road mostly sloshy,
a full moon illuminating big smoke rising
from what must have been pit or ditch.
Three dark shapes nosed. Eyes flared,
quick incandescence amidst rubble.
They're eating America, I joked,
though no one laughed and it wasn't funny,
this view of large shadow creatures foraging
the land uphill from a small fishing town
accessible only by small plane or boat.
We sat silent then, until we too had our fill.

Nick Jans

Crossing Paths

I SAW HIS SIGNS FOR WEEKS before I met him. I'd wake up in the morning and there they'd be—fox prints, dainty and catlike, right outside my door. If I'd been careless enough to leave out a trash bag, scraps fluttered in the wind. The red fox must have passed almost nightly, sometimes hunting for voles around my woodpile, leaving pounce marks in the snow. But though I watched for days at twilight, shone flashlight beams from my windows, and made quiet stalks home in the afternoon, I never caught a glimpse of the ghost that floated through my yard.

Then, one day, there he was, trotting close to the tree line in bright sunlight, bold and unconcerned, headed right toward me. Caught in the open, I froze. The fox, a red orange male with black forelegs, kept coming, finally passing within twenty feet, glancing my way just enough to leave no doubt: he'd known I was there all along.

After that, I saw him almost every day for several weeks. I'm sure it wasn't coincidence; the fox had somehow decided that I was safe, and only then did he allow himself to be seen. I began leaving tidbits for him, and within two weeks, he allowed me to approach within several paces. While still wild and shy, he sometimes didn't flee, and sat with his tail curled over his toes, head cocked, regarding me. We each knew something unusual had passed between us. I resisted giving him a name, but I began to look for him, and missed him when he didn't show.

Some of my most memorable Alaska wildlife experiences have occurred far back in the country, up nameless canyons and on tundra plains far from any settlement. But the dozens of encounters I've had in the shadow of my house, both in Ambler and in Southeast, count among the richest. Knowing that wild creatures move in the space I call home reminds me of my connection with their world, and why I'm here. The mountains and rivers and glaciers of Alaska are stunning, but I was lured here by something else—the countless living things that flow and breathe across the land's face.

When we think of Alaska wildlife, we think of big stuff—moose, bears, caribou, and wolves. They thrive in valleys that seldom see human passing, but certainly all of these creatures also drift, without our bidding, through spaces we consider as ours. Black bears stroll through downtown Juneau, right past the governor's mansion; a pack of wolves hunts moose and house pets through sections of suburban backyards in Anchorage. It makes sense when you remember that ninety-odd percent of the state's expanse remains scarcely brushed by the hand of man; wild living things, large and small, are bound to cross paths with us.

Ambler, my home village for two decades, is tucked above the Arctic Circle on the southern flank of the Brooks Range, far from the road system. No surprise that lynx padded through the willows on the edge of town, and that the gravel runway could be clogged with caribou in September. I lived among hunter-gatherers, and yet there were often wolverine or brown bear tracks within a half mile of my door. I chased weasels and squirrels out of my *qanisak* (storm shed) with a broom, and once had to swat away an owl as he dove toward my face, drawn, it seemed, by my fur cap. Once I reached out and touched a ptarmigan nestled in the snow by my shed, and another time sat drinking my morning coffee among a family of snowshoe hares.

While most Alaskans think of themselves as loving nature, many draw the line when it comes to their backyards. It's an issue of safety, they argue; kids and pets don't mix with moose, and trash-rummaging bears are a menace. They recall incidents—that old guy stomped to death at the door of a University of Alaska building, an Eskimo man killed by a polar bear right outside his house, and countless near brushes every year, with critters ranging from porcupines to wolves.

They have a point, no doubt; wild creatures can be unpredictable. But when you consider the number of animals each year that are killed in what's officially called "defense of life and property" and weigh them against a spare handful of human casualties—few of them serious—within city limits statewide, it's pretty clear who's dangerous to whom. Then there are the hundreds of moose slaughtered by trains and cars, and the bald eagles electrocuted on high-voltage lines. As a species, we have a sad, inevitable history: wherever we choose to live, we displace—either by necessity or choice—the wild. In the cold terms of logic, alternates exclude.

Yet, by world standards, Alaska represents one of the great oases of wilderness anywhere, a rich, sprawling landscape where entire ecosystems have remained intact, large predators and all, across centuries. Maybe that's why I sit in Hoonah, where brown bears sometimes roam across the school playground just two hundred yards from my door, and feel discouraged. Alaska is our last chance to somehow learn to coexist with the wild, and we seem to be reliving, one death at a time, our past and our entire westering progress across the continent. My friend Beth up in Ambler felt she had to shoot two black bears in her yard this summer, mere feet from her door (never mind that the bears were drawn to piles of unsecured, aromatic dog food). That was just one person in a tiny village. How many other creatures did a half million other Alaskans kill, not for food or even sport, but by accident, or because they felt there was no choice?

I remember the twinge of despair I felt as I held a tiny, rare boreal owl in my palm, its neck broken against my picture window. The bird died because, by building a house, I moved into its space, not it into mine—a small distinction, perhaps, but a telling one. And maybe this was just a small death, one that most people would shrug off, say it couldn't be helped, and no big deal. But what, finally, is the difference between a bear and a robin-sized owl? Wherever we choose to live, sooner or later other species can't.

Maybe we don't outright kill everything. We set out birdseed and hay bales, leave sheltering stands of trees or brush, set aside parks and reserves. Alaskans, more than most Americans, accept, even embrace, the presence of wild creatures in their lives. But parking lots, trailer parks, and condo developments don't cut wild things much slack. They usually have to give up that habitat and move on; even if they could stay there unharmed, there's not enough to hold them there.

What of the red fox I befriended? Well, he didn't show for one week, then two; I decided he must have moved on. But I found, by chance, that a neighbor and friend of mine had shot him. The fox had been hanging around, a little too close for comfort. Unafraid. Very odd—the same way a rabid animal might behave. My neighbor had a dog, and grandchildren who sometimes played in the yard. Taking out the fox was an understandable precaution. And so, in a sense, I pulled the trigger. If I'd thrown rocks instead of caribou bones, the animal would have known better. Being at ease around our kind was a luxury he couldn't afford.

Twenty years ago, Munz Airlines in Kotzebue coined a slogan—See the Arctic Before It's Paved. We're a long way from that actuality, but there's a kernel of truth nestled beneath the hyperbole. Once upon a time, grizzlies roamed the hills around San Francisco, and bison galloped across western Kentucky. Surely we've come far since the days of willful extermination, but the essential truth remains: people take up space, and there are more and more of us each year. Even Ambler, tucked back in a remote corner even by Alaskan standards, has doubled its footprint since I first saw it. No doubt there will come a time when a road will snake its way north to the Kobuk valley, and why should our all-consuming progress, for once, stop there?

Still, when I returned for a visit to Ambler this summer, I found a big snowshoe hare had set up home base under my house. Moose tracks wound down the main trail. I saw three black bears in midafternoon, minutes from my door. I say these things, hoping for reassurance; maybe, just maybe, I tell myself, there will always be fox tracks by the woodpile. The heart is, after all, a resilient little muscle.

Tom Sexton

Sweet Spring Grasses

Trophy hunters killed the cinnamon-
hued bear that haunted your tundra lake
days ago as it ambled from the willows.
Skinned and measured, it lies like an obscene
Buddha toppled in its blood-stained clearing.
Its flesh sours. Maggots bubble in its fat.
Dreaming of salmon and sweet spring grasses,
it must have wintered in the nearby hills.
I follow the ridgeline to our cabin.
Not even Denali, floating above
the kindled mist like a sacred mountain
in a Chinese painting, can lift my spirits.
As I cross an open meadow, light, like
small flecks of shattered bone, falls on my hands.

Anne Coray

Precarious Preserve

AMONG OUR OLD FAMILY PHOTOGRAPHS is one taken in the spring of 1959. My brothers, Paul and Craig, stand on the far right. Not yet teenagers, they are dressed in typical Alaskan clothing: flannel shirts and jeans, Paul in shoepacks, Craig in hip boots. Patches of snow still cover the ground. Most of the picture is dominated by our woodshed, a building of rough-cut spruce that long ago rotted and collapsed. But in the photo the woodshed is sturdy, and tacked to its wall are the pelts of several animals: six foxes, three lynx, and a wolverine. My brothers are smiling. By trapping standards this is not a large catch, but it is a source of pride for two youngsters who, at this formative period of their lives, measure their worth against the trades and practices of men around them.

I feel nostalgia for the old days when I look at this picture. I feel regret, because Paul is dead now, and this period was one of the happiest in his life. I feel strangely detached from the pelts—which no longer represent living, breathing, animals—and my response to them is merely intellectual: I know that their lives were taken so that my brothers could earn a few dollars. Trapping was little more than a hobby for them, as it is for many today, though some people in rural Alaska still use it to supplement their income.

Decades ago, trapping was a viable occupation. Our neighbor, Brown Carlson, ran a one-hundred-mile trapline from his cabin east into Little Lake Clark and west as far as the Kijik River. One year in the 1950s he earned over $3,000. He traveled by foot, usually on snowshoes, using pack dogs to help transport his supplies.

Where I live in southwest Alaska, within Lake Clark National Park and Preserve, trapping is still legal for residents. In historic terms, the park is relatively new. Established in 1980, it must follow mandates of the Alaska National Interest Lands Conservation Act (ANILCA), which both protects land and makes allowances for customary usage activities. Consequently, trapping is considered a means of livelihood/subsistence, and snowmachines a traditional means of transportation, though they only became popular in the 1960s.

In the Lake Clark area, the Park Service does not establish its own trapping regulations, but adopts the state's, meaning that no limit is set on the number of animals that may be trapped within the designated season for each species. These seasons are liberal, typically beginning in October and lasting at least through February, often into March or April. As a result, the pressure on furbearers subsides only when weather makes travel difficult.

Recent winters have been mild in southwest Alaska—abnormally so. My husband, Steve, and I bemoaned the December rains, the forty-degree-above-zero January, the open water. We longed for solid ice, which provides a smooth, safe runway of unlimited length for Steve to land our Super Cub. We wanted quick, easy access by snowmachine to visit our neighbors, the Hammonds, five miles distant, or their winter caretakers, Tom and Sue, with whom we've become friends.

In November 2003, though, the temperature dropped, hovering in the teens and twenties. December brought single-digit readings and skim ice in our bay. By January the lake had completely capped over. We were thrilled. It felt like *Alaska*.

One day, skiing out to a point from which we have a nice view of the upper end of the lake, we noticed a black speck on the opposite shore, four miles away. A wolf? Too big. A bear? Not in the middle of winter. Then it began moving. It was a snowmachine. Good travel for us meant good travel for others. In this case, a trapper.

Later, we would learn who this person was, that his trapline ran from Port Alsworth, located near the Tanalian River on Lake Clark, up the south shore of the lake as far as Currant Creek, a distance of roughly twelve miles. And he wasn't the only one. Another man claimed the remaining south shore, circling into Little Lake Clark, then down the north shore as far as "the cove," where a number of summer residents have cabins. And another—the new neighbor between our place and the Hammonds'—had begun trapping on his 160-acre homestead. And others—farther down the lake's north shore, from Kijik River on into the village of Nondalton. Our stretch of beach was one of the few on the forty-five-mile-long lake that supported no trapping activity.

I shouldn't judge—or should I? Something seems morally amiss when trapping within the park itself is usually carried out by residents of Port Alsworth, a predominantly White community in a high income bracket with a year-round population of roughly one hundred people. Port Alsworth, like Nondalton, is classified as a "resident zone community," meaning that any state resident who moves there automatically qualifies for the same hunting/trapping privileges, regardless of longevity in the area. (By contrast, Nondalton is composed primarily of Denai'na Athabascans, whose generational and traditional use of resources is more in keeping with my interpretation of ANILCA's intent.)

Of course, Steve and I are grateful that people trapping at the upper end of the lake seem to respect our presence. We have a good six or seven miles of shoreline to walk, skate, or ski without worrying that our dog, Zipper, will be caught. Without having to witness firsthand an activity we look on with increasing disfavor.

Recently, I've felt compelled to learn more about the methods and means of trapping. The most "humane" trap is the killer-style trap, such as the Conibear, that crushes the body of an animal walking into it. These traps come in different sizes, but they are not well suited to animals like foxes, coyotes, or wolves that don't like to venture into confined areas. Occasional damage to the fur occurs, and more disguise of the trap is necessary than with other types. Nor do they always work as advertised. My friend Jack Ross tells of a lynx he trapped with a Conibear. When he found it, the animal was still wheezing, metal cinched around its lungs.

Leg-hold traps—single, double, and coil spring—are commonly used for many furbearers, especially fox. Snares are also popular. Both snares and leg holds are often placed near trails or kills in what might be called "carpet bombing"; that is, enough are set in a given area that an animal may be caught by two, sometimes three, legs.

Beaver and occasionally muskrat are trapped underwater, where they drown. This is known as the under-ice set. The range of sets for various species seems inexhaustible, and even a member of the Society for the Prevention of

Cruelty to Animals might marvel at the creative energy expended in the trapper's efforts to conceal and deceive: the tunnel set, the pocket set, the cache set, the buried bait set, the stove pipe and hollow log sets, the artificial feed set, the castle set, the tree stump set, the curiosity set.

Who doesn't long for a quick death? Who doesn't believe that suffering—whether human or animal—is our least desirable exit? I cringe when I hear horror stories of wolverines, which will sometimes chew off a trapped foot in a last desperate effort to escape. Maimed, those animals' chances of procuring food are slim, and thus so are their chances of survival.

One January, Paul, a visitor at the Hammonds' homestead, came upon a fox with one bloody stump, dragging a trap with another paw. With no weapon, Paul was unable to dispatch the animal. It snarled at him before limping awkwardly into the woods.

Jack Ross tells me about a fox he found that had been caught by its jaw. He doesn't go into detail, only says that this was one of several incidents that gave him a change of heart, so that he eventually gave up trapping altogether. "I just couldn't do it anymore," he says, shaking his shaggy eighty-two-year-old head. Some locals would say Jack has "gone soft."

Modern-day trapping raises other ethical concerns. Like commercial fishing, the equipment that provides access to the resource has become so efficient that it is possible to trap animals out of an area, just as it is possible to wipe out fish from a sea. In Bristol Bay, double-enders with sails have given way to boats sporting sonar, radar, and GPS, powered by twin diesel engines. Throughout most of bush Alaska, working dog teams have all but vanished. Relatively lightweight single-cylinder snowmachines were the first replacements. These metamorphosed into six-hundred-pound hydraulic-suspension behemoths with heated handle grip and adjustable backrest options. Tradition dies hard, but we accept with alacrity advances in technology.

I worry about the lack of bag limits. While the trapping community is quick to make assurances that no furbearers in Alaska are threatened or endangered, there is no management tool other than harvest records by which the Alaska Board of Game sets regulations. (Trappers are required to seal river otter, lynx, wolf, and wolverine, as well as marten and beaver in most game units, and it is on these returns that regulations are based.) In *The Stars, the Snow, the Fire* John Haines writes, "It is possible to trap every marten and mink in the country, and the same will be true for lynx. A country trapped too hard, as my own Redmond creeks and hills had once been, may take a long time to recover, and a man living there will face lean years of his own making."

Nor is there a plan to manage the human population within Lake Clark National Park and Preserve. This is a concern. I've always believed that game management and discussions of sustainability should include proposals for balancing the human population as well, but this is still considered a radical idea in twenty-first-century America.

On trips to Port Alsworth to retrieve our mail, Steve has observed men with children in tow, making preparations to track down wolves. One little boy four or five years of age proudly carried a popgun. A girl, eleven or twelve years old, packing a .22 rifle, tagged along with her father, who had a semiautomatic assault-style weapon slung across his back and announced, "We're going to slaughter some wolves." Such anti-predator convictions are common here, and are generally shared by hunters and trappers alike.

How can a change in consciousness begin to find footing when on the front cover of the 2003–2004 Alaska Trapping Regulation Booklet there is a three-year-old boy posing with a toy gun and the body of a lynx? The lynx, not yet skinned, hangs by its hind feet from a tree. It resembles a huge, furry stuffed animal, the kind a little boy might ask his mother to fetch before he is tucked in bed for the night.

The fur looks soft. So soft that I am compelled to leave my writing, walk to the back wall of our cabin and take down the red fox pelt that hangs there. A friend of Steve's trapped this fox years ago, and it serves to remind me why there is still a market for fur despite the controversy surrounding trapping.

I wrap it around my neck. It's plush, warm, clean, and beautiful. It smells faintly of tanned hide, a pleasant and earthy odor. I imagine how luxuriant it would feel to wear a full fox parka, and the aristocratic mystique I would project in such a garment. It is too easy to separate the skin from the living animal, and I see how readily people fall victim to fashion. But that sense of enchantment soon fades.

I think of the delight I find in watching foxes travel through our yard, or the hours Steve and I spent watching one from our cabin window. It hung around for days, contented and unthreatened, returning our gaze with its dark, liquid eyes. It pounced on voles and looped its tail to form a pillow whenever it lay down to rest on the snow. There is no photograph so lovely.

I'm especially fond of foxes. That's why I felt heartsick at the sight of the carcass, about the size of a Scottish terrier, in our neighbor's yard. The fox had been flung just outside the door and had apparently been skinned some time ago, for the flesh was dark and hardened. A clean knife cut left four little boots on the paws, probably for ease of skinning. I don't know what it was exactly—the stark contrast perhaps between the handsome, vibrant animals I had been lucky enough to see up close and the meager corpse exposed haphazardly to the elements. I felt an immediate, visceral revulsion at the sight of it and quickly glanced away.

Though it may appear contradictory, I am not opposed to hunting. I support the consumption of wild meat because is it healthier—free of hormones, antibiotics, and excess fat—and it does not require the destruction of forest as does the raising of grains or livestock. I have thought deeply about the hidden costs of vegetarianism: the by-catch of birds and rabbits as combines sweep through the fields; the transport, with fossil fuels, of corn, wheat, and rice. Yet I never enjoy killing animals or watching them be killed. And every moose I have helped butcher has elicited that same sense of misgiving and loss. It does not seem fair that nature insists on the exchange of life for life.

My friend Sue asked me recently, "How is it that you are such an environmentalist, being born and raised in Alaska?" I paused. I wasn't particularly struck by the presumption that Alaskans as a group have few conservationist leanings. I'd heard that often enough. But I hadn't really thought about what had helped shaped my views. "You know," I said, "it's mostly from reading, and thinking about things." Until that moment, the impact that reading has on me hadn't fully registered.

I recognize that there are responsible trappers who check their lines every day to reduce the amount of time an animal spends suffering. Some elect to shut down early in the season, if sign of a given species is diminishing, for they are interested above all in perpetuating the resource.

I admire women who practice the art of skin sewing, as I admire the tradition from which it arose. The garments produced can be beautiful and are often executed with amazing skill. I even understand how tourists are drawn to souvenirs made from scraps of fur, though I am not a collector of knickknacks. There are two fur hats in my house, and I occasionally wear one of them. But I would not purchase such an item today, when synthetic materials are perfectly comfortable and provide adequate warmth.

I didn't grow up with trapping—I was too young when my brothers were actively involved—but the stories are there, and the family photographs, so that in many ways it does feel like part of my heritage. It's not an easy thing to completely denounce. With stricter regulations, I might find it easier to live with. Still, I wonder what are we preserving in this park and preserve—wildlife, or a human lifestyle that relies too heavily on modern technology?

Our footing in the history of environmental awareness is still precarious, an adjective whose root is *precarius*, from the Latin, "dependent on prayer." But prayer, for me, has proven ineffectual. I recall the inner pleas I have made to wolves, foxes, and wolverines passing in front of my cabin: *Don't go that direction; turn around; it's safer back up the game trail.* No matter how fervent my request, it never altered their course.

John Haines

The Man Who Skins Animals

I

He comes down from the hill
just at dusk, with a faint
clinking of chains.

He speaks to no one, and when
he sits down by the fire
his eyes, staring into the
shadows, have a light like drops
of blood in the snow.

II

There is a small, soft thing
in the snow, and its ears
are beginning to freeze.

Its eyes are bright, but
what they see is not this world
but some other place
where the wind, warm and
well fed, sleeps
on a deep, calm water.

Eva Saulitis

Wondering Where the Whales Are

But what surface have we fallen through,
Here beneath the trees? What do we see in our infinity
if each is all the same, or all unknown?
—Molly Lou Freeman

HALFWAY DOWN AN ALDER SLOPE, through an opening in the jumble of branches and leaves, Craig's son, Lars, spots them. "Look, killer whales! They're coming into the bay!" Even though he's only eight, Lars knows whales from a mile away and five hundred feet up. Since he was six months old, he's been bundled and stashed on his parents' research skiffs every summer. At "Whale Camp," he rubbed holes in his baby socks bouncing on the wall tent's plywood floor, his Johnny Jump-Up suspended from the ridgepole. "Listen," he says. "You can even *hear* them. They're over *there*." He points; we squint. There, in late ponds of sunlight, we see black fins and breath smoke.

"Good eyes, Lars," says Craig. "Let's go." We lurch down the mountainside, branches tearing at our hair and our faces.

On the beach, we gallop, digging wedges into sand with our rubber boots. In the distance, a few yards off shore, the first bubble cloud rises, and we run for it. A fin slits shadows at the sea edge. When we get closer, we tiptoe. Five killer whales slide along shore, releasing air so they can sink. White bubble-rings bloom and phosphoresce on the water's surface. Beach-rubbing, it's called: on beaches with particular slopes and small, round pebbles, killer whales approach shore to slide their bodies along the bottom.

Craig and I, unthinking, pull off our boots and socks, make to unbutton our shirts and jeans, to jump into the water, where the whales are. Lars asks, *"What are you guys doing?"* and we look at each other and laugh. What *are* we doing? We fall heavily onto the beach, side by side, and watch. The twenty feet separating the whales from us might as well be the Gulf of Alaska. We can't cross it.

For thirteen years, we've studied killer whales in Prince William Sound and Kenai Fjords. Fishermen still don't quite understand what we do out here. They call us "whale-watchers."

Listen, I want to say, we're not *tourists*. We're doing *research*.

But we do; we watch. We watch from shore, with our boots askew on the ground; we watch from the boat's deck, poised with our notebooks, pencils, cameras, binoculars, with vials in which we place a sample of whale skin. The whales visit our dreams, where they watch us. So what are we to be called—scientists, observers, natural historians, writers, intruders, watchers? The killer whales are called *aaxlu, takxukuak, agliuk, mesungesak, polossatik, skana, keet*, feared one, grampus, blackfish, orca, big-fin, fat-chopper. In the language of science, they are *Orcinus orca:* whales from the realm of the dead.

In grad school, listening to a lecture, I stare out the window and scribble along the margins of my notebook pages:

outside
birch fingers cast smoke
ribbons on snow

The professor chalks formulas on the board, flips on the overhead projector's light, casts a graph on the wall of the oxygen consumption of a marine mammal. *The language is like this*, I write in my notebook:

kinetic
isotope
extrapolate
index

The formula makes sense. Someone has figured this out, disproved an old theory about the way whales dive. I memorize the formula, stash it like some possibly useful thing into that catch-all drawer in my mind.

In another part of my mind, in a dream, I'm standing on the *Lucky Star's* stern with Lars. In the boat's wake, a pod of beaked whales circles. They twist in the foam; their eyes glint up. When Lars reaches out his hand, a whale grabs it and won't let go. Using all of my strength, I yank him free.

When I told a Yup'ik woman about this dream, she called it a warning. In her Bering Sea village, she was taught that killer whales must never be hunted or bothered. We must not touch them.

But we want to know.

We want to know, like the raven, who, in a Chugachmiut legend, swims into a killer whale's mouth. An old woman sits inside the whale and asks, "How did you come here?" The raven answers, "I called him. I wanted to know what was inside him."

For several summers, I worked out of a tent camp and ran a small boat with friends who volunteered to be my field partners. The first summer, I needed to decide on a project for my master's thesis. We followed two kinds of killer whales in Prince William Sound—fish eaters and mammal eaters. Little was known of the behavior of the mammal eaters. Seldom seen, shy, difficult to follow from a boat, they were called transients. A group of twenty-two transients Craig named the AT1 group belied their name; they stayed mainly in Prince William Sound. I grew curious about how they communicated, since transients are almost always silent.

One day, my friend and I followed two AT1 transients as they hunted harbor seals along an island shore. We lost them for several minutes, and then spotted silver mist above a rock. We let the boat drift near. Clinging tightly to the rock, its head craned back, eyes huge and black, a seal pup crouched above the water line. A transient nudged the rock, but couldn't reach the seal. Abruptly, the whale turned, joined the second whale and swam rapidly across an open passage. We left the seal and raced to catch the transients, but they'd vanished. Cutting the engine in mid-passage so we might hear their blows, we stood up, scanning with binoculars.

It seems I felt something through the bottom of my feet before I heard it. From the boat's wooden floorboards, a wail rose, and another, and another. My friend and I stared at each other.

"It's the whales. They must be right under us. Let's drop the hydrophone," I said.

I scrambled for the tape recorder, and we huddled over the small speaker as long, descending, siren-like cries reverberated against underwater island walls. In the distance, other whales answered, faintly.

I'd never heard transients call before; it was like a stone had sung. I knew then. I wanted to learn the language of the whales that were mostly silent.

In grad school, I learned the art of detachment. I learned to watch how I said things, to listen for things "anthropomorphic," like applying the word *language* to non-humans. As scientists, we distinguish ourselves from whale huggers, lovers, groupies, and gurus, from those who think of whales as spiritual beings. We learn the evolutionary, biological basis for an animal's behavior. We learn the various theories and counter-theories and debate their merits: reciprocal altruism, game theory, optimality theory, cost-benefit analysis. At scientific meetings, in animal behavior seminars, we don't debate whether animals have feelings. It is *terra incognita*. But on the research boat, at breakfast table before the meeting begins, some of us argue these things. Or non-scientist friends, puzzled by the ways of science, ask, if it is anthropomorphic to suppose that animals might share qualities with humans, then what are we? What kind of species? Is it "animapomorphic" to ascribe animal traits to humans? And what is it—what trick of logic—to assume that only humans have feelings?

Out in the field, for nearly twenty summers, I search for knowledge. I practice the method of science: observation, hypothesis, data collection, analysis, discussion, conclusion. Poet and biologist Forrest Gander says that this method "has endured as a scientific model, and a very successful one, for it predicts that when we do something, we will obtain certain results. But . . . if we approach with a different model, we will ask different questions." To create a new model: that prospect challenges all of the questions I've learned to ask.

Over the course of a four-month-long field season, sometimes we see killer whales every day, and sometimes weeks go by without them. Often, we've spotted distant whales, come near them to take photographs, and they've vanished.

In response, I do things that annoy my field partners. Like a mariner who touches wood for luck, I've developed irrational habits. When I spot distant whales, I leave the camera in its case. I don't jot notes in my field notebook. I don't get ready until the whales are beside the boat, so I don't presume. Writers like anthropologist Richard Nelson and Yup'ik scholar Oscar Kowagly teach me that in the worldview of Koyukon Athabascan and Yup'ik hunters, arrogance and presumption lead to bad luck with animals. And after twenty years, finding and staying with transient killer whales sometimes feels like one part logic, two parts eyes, two parts knowledge, and nine parts luck.

On the luckiest of days, we've followed killer whales for twenty-four hours. Then, after so long, even the observing eye becomes insufficient, so we listen. In darkness, we navigate by the sound of their breathing.

We take turns sleeping. I'm leaning on the boat's dashboard, echo of engine roar dying, wave slaps against the boat's hull taking over; I drop the hydrophone down; I listen. Two in the morning, just past summer solstice, Montague Strait is dimly lit, but it's too dark to see the whales. I don't hear anything. They're down for a long dive, or we've lost them. I hold my breath. Then, a few hundred yards away,

whoosh-ah whoosh whoosh whoosh-ah

I stare in the direction of silence where the sound was, hear water closing around it.

Another summer. For eleven days, we search hundreds of miles without finding killer whales. On the twelfth day, we hear radio reports of a large group forty miles from camp, in open water. After roaring past Smith Island and Little Smith Island at twenty knots, we drop the hydrophone, climb onto the cabin, scan with binoculars, radio the boats that reported the whales. The water's still. No one responds. On the hydrophone, waves: lap, lap, lap.

After searching an hour, we give up, then devise a plan. No more running around wasting fuel. We're going to wait, let the whales come to us. We return to camp, and the next morning, we gather paper and books, food and a thermos. We run the boat a mile off camp, out in the passage, and shut down. After dropping the hydrophone and scanning around, putting out a radio call, we settle into the boat, build a fire in its tiny stove. Every half-hour, one of us pulls on raingear and climbs onto the roof and scans with binoculars.

A couple of hours pass this way. I look at my watch, put down my book. It's my turn to scan. I glance out as I reach for my jacket. Fins rise around the boat.

It worked! I shout, and we rush to put on rain gear, gather cameras. The rest of the day, we follow the whales.

The same trick never works again. We're constantly second-guessing. Should we sit still? We call that the "sit and wait" hypothesis. Should we move? We call that the "Lance and Kathy" method, after colleagues who averaged a hundred miles a day one summer. Once, Craig and I searched the outer coast of Montague Island, over seventy miles, and saw nothing. Our friend radioed that killer whales were off Point Helen, a few miles from Whale Camp. If we could, like Fitzcaraldo, drag our boat up over that mountain, we'd be right there.

That night, we watched a wildlife DVD about a group of filmmakers who spent four years trying to photograph snow leopards in the Indian Himalaya. They never saw a kill, never saw a litter of cubs, their two greatest desires. They didn't see snow leopards for the first several weeks, just tracks and scat. They concentrated on these tidbits, mapping them until they sensed patterns. Even after twenty years in the field studying these creatures, we have to revamp our intentions and strategies: to concentrate, not on our desires, not on the past, but on clues, which is hard when you're lifting fluke prints off water.

The morning after watching the film, three AT1 transients slid past our anchorage. We heard them first on the hydrophone. They were hunting marine mammals offshore, diving for ten minutes at a time, constantly changing direction. The north wind blew up. We lost them after an hour. We managed to take one identification photo.

In the Tlingit language, the word for killer whale, *keet*, means *supernatural being*. But how reliable or subtle is this translation? In nature, creatures defy our predictions. In the 1980s, biologists divided fish-eating killer whales into pods, extended family groups that remained together for life. Recently, that story has been revised. These societies orbit around the matriline, mothers and offspring, and pods sometimes fracture for unknown reasons. The loss of a key female may cause a family to rupture, bonds to loosen. Discoveries only add to the complexity, to the *keet* nature of the wild animal. And the more we know, the longer we stay, the more we care, and caring, like anthropomorphism, is tricky ground for that detached creature, the scientist.

For the past few years, we've been collecting tissue samples from killer whales to monitor contaminant levels in their blubber, to extract DNA from their skin. We've found out that their breeding populations are small, a few hundred animals, so an oil spill, a die-off of salmon or seals, can be catastrophic. From genetic analysis of the samples, we've found out that killer whales live in specialized societies: mammal hunters, or transients, and fish eaters, or residents. The two types don't interbreed, though they share the same waters. We've learned that transients carry high PCB and DDT levels in their blubber, that mothers pass these toxins to calves through their milk. But to learn this, we must approach whales more closely than we do to take photographs. To do this, we point a rifle at a whale and shoot a biopsy dart into its body; it pops out after catching an inch-long piece of flesh on its thread-like barb, and we retrieve it from the water with a dip net. To do this, Craig and I argue with each other. We experience conflicting feelings. *We can't dart now; they're resting. These animals are rare. We can't dart in front of tour boats. We might not have another chance. We've probably darted enough animals in this group. We need more samples for the statistical tests. We have to have a common mind. I hate all this.*

Even Lars, who's enthusiastic about shooting, scrunches down in the bow, fingers plugged in his ears, eyes shut tight when the shot's fired.

From the boat's cabin top, I scan Montague Strait in light diffused by high clouds, looking for blows. I spot a white glittering, then another. It's the kind of haze made by a leaping whale when its body collapses onto the water.

We race that way and find killer whales, take identification pictures of their dorsal fins and flanks. They're Gulf of Alaska transients; the last time we saw them was four years ago, and we don't have DNA samples. Their calls are different from those of the local AT1 transients, so they might be from a completely separate population. Today, Craig isn't here to wield the dart gun. Today, my field assistant—my husband, John—and I have to do it ourselves.

For the next two hours, the whales lead us past Danger Island, farther into the Gulf. John, more comfortable with a rifle than I am from his years in the Alaskan bush, shoots five times without success. Out of Montague Strait's strong current, the water calms to a swell. In exasperation, I've taken the gun. John pulls the boat in close to the whales, and I site on an old female's scarred saddle patch. Without thinking, I pull the trigger. The dart hits the saddle patch and bounces out. She slaps her tail and dives.

"We got a sample," I shout, elated, when I pull up the dart and see blubber protruding from the tip. I give John the gun on the next approach, and he darts another female.

"We're getting pretty far out here, " he says, after I wrap the third sample in foil and store it in the cooler. "I think we should go back." I glance toward Prince William Sound. We're at least four miles from shore now, and the whales are heading steadily south in the direction of Hawaii. As we drift, we watch them disappear.

An hour later, we're anchored up at Foxfarm Bay, just inside Cape Elrington, processing samples. Intent on my work, thrilled at our success, I don't notice John watching me.

"I've never seen you that way before," he says.

"What do you mean?" I say, looking up.

"You were so angry and impatient, even rude at times, and then, suddenly, when you got the samples, you were ecstatic. A real Dr. Jeckyl/Mr. Hyde. It was scary."

I stare across the bay, where a sea otter lazily rolls and dives and brings up some kind of shellfish. Inside me, a nauseous feeling rises.

I haven't darted many killer whales since. It's Craig who wields the gun. And there are whales we've never been able to dart, mostly sea lion hunters with torn fins. They sometimes approach our boat, curious, staring at us with inscrutable eyes. Once, a female grazed her body along the skiff's side, her mouth open, showing rows of perfect teeth. "What are you saying?" I called after her as she swam away.

Years ago, another whale drifted under the bow where I stood, looking down. She held a harbor seal in her jaws. Blood from the seal's body throbbed.

Science trains me to be detached in moments like these, but I'm sometimes angry or panicked in the field, when I can't get what I want, what I *must* have. I have no control over what's invisible, that which binds me so viscerally to my desires, and that which decides when the whales will find me.

After several days without whales in Resurrection Bay, a fjord west of the sound, Craig and I overhear a radio conversation between tour boats. Killer whales are traveling along the rocky shoreline of Fox Island, fifteen miles from where we're floating, our hydrophone down. They're heading for the cape, out of the bay and out to sea. The skippers think they're transients—the ones they call "the bad boys"—two large males that hunt harbor seals in ice floes off the Aialik Glacier.

We drop our books and scramble to start the engine. We call a skipper, get a location and direction of travel, and we roar across the bay, coaxing as much speed as we can out of the *Whale 2*. When we spot the whales, we know right away they're not the local "bad boys." Their fins are too broad and tall. As I slide the boat in parallel to the whales so we can take pictures, I scan photos of transient dorsal fins in the killer whale catalogue.

"Who do you think they are?" Craig asks, clicking off frames. "They're awfully tolerant for Gulf of Alaska transients."

The whales seem indifferent, slowly traveling, breathing for eight breaths, then diving for ten minutes. They follow a regular compass heading east, directly past Cape Resurrection, toward the sound. I stare at two blurry photos, then back up at the whales.

"They *are* Gulf of Alaska transients. They're the AT30s." The pictures are poor. They've only been seen once before, seven years ago, in bad weather.

We spend the next hour trying to get biopsy samples. Tour boats come to watch them, so we don't dart. Darts miss. Once, a dart pops out of a whale but doesn't take a sample. Another time, we're too far away when they surface. Other times, they change direction slightly when they dive. I plead to them, to Craig's amusement, as I position the boat. "Whales, please let us take these tiny samples. We'll never have to do this again. It's for your own good!"

We call out names for them, Chubby Rain and Heavy Rain. Despite our blundering, our absurd behavior, the whales let us approach closely again and again, and finally we have some samples.

Floating off Killer Bay, we watch them disappear. "Don't you wish you knew where they were going?" Craig asks. "Someday, with a little transmitter attached to them, we won't have to wonder where they are."

Now I can barely make out two distant, black triangles among rolling hills of water, and I think of them unwatched by anyone for eight more years. They're swimming off the edge of the known world, like hapless ships on ancient charts. They might dive right through the sea realm, resurface in some other, a realm of the dead.

A young Sugpiaq man from Nanwalek, in outer Cook Inlet, told me there's a lake near his village that's bottomless. A killer whale jumped into that lake, he said, dove to the bottom, pushed through and emerged in another lake.

We cling to what we know. In response to Descartes' mechanistic view of the universe, Blaise Pascal said: "The silence of these infinite spaces terrifies me."

Science. It seems solid, but it's mostly space, like a gill net I drop over the world. Two transients pass through its web, leave me holding a tiny sample, a pencil.

A young scientist seeks mentors. I've had many: Bud Fay, my major professor in grad school, Craig, other whale biologists, and those I know through their discoveries, their tenacity, their eyes that see what others miss, my biologist heroes. I met Mike, my last mentor, one afternoon at Chenega Bay, a Chugachmiut village in a remote corner of Prince William Sound. He rode his four-wheeler down the steep ramp to the dock. In the vibrating silence after he'd shut down the engine, he sat and watched me as I pumped fuel onto the *Whale 1.* His look was inscrutable. There was no smile; under his cap, his eyes were shadowed. He could have been angry. Non-Natives were not always welcomed in the village. I tensed when he climbed off the four-wheeler and, hands in pockets, strolled over to the boat. "Seen any whales?" he asked, grinning.

He was all sinew, brown skin, black hair, and a small, bowlegged frame. He wore a plaid wool shirt, stiff, new dungarees, and wire-framed glasses. I knew he was considered a village elder, although I couldn't tell his age. He coughed often, into his fist, turning his head away. I introduced myself, but afterwards, he'd show up at the dock whenever I was there and greet me, "Hey, *Whale 1.*"

Mike was born on Chenega Island, at the old village. During the 1964 earthquake, a tsunami destroyed the place. The survivors abandoned it, rebuilt here on this island, twenty miles away from the original village. Mike was one of the survivors. Now he was dying of lung cancer. He dropped bits of knowledge into our conversations: where he'd seen whales, how seals in the area were declining. I knew he hunted seals but learned only from other villagers that he was one of the most respected hunters in the sound and one of the last in his village.

I began to look for Mike when I came to Chenega Bay, wandering to his house, inviting myself in for a cup of tea. Somehow, I felt attached to him. Our conversations were brief. But, after time, when he saw me, he hugged me. He teased me. When I told him what I wanted to be, he shook his head. "Why do you need to do that? You don't need to go to school to do that. You just need to live out here."

The smell of burning alder drifts up from Mike's smokehouse. He's gone today. He's hunting seals.

We anchor the boat in front of camp. It's sunny, but the wind's come up, so we decide to take turns trying on the dry suit, snorkel, and mask and swimming through the eelgrass and kelp beds. My friend helps me with the zipper.

> *I put on*
> *the body armour of black rubber*
> *the absurd flippers*
> *the grave and awkward mask.*

I hear words from Adrienne Rich's poem in my head when I drop feetfirst from the boat's side into the sea:

> *there is no one*
> *to tell me when the ocean*
> *will begin.*

After I pull the black rubber away from my neck to release air, the dry suit clings to my body like loose skin. I place my face in the water and breathe through the snorkel, wheezing rapidly at first out of fear, and the sound is loud, like the breaths of someone dying.

Eelgrass and kelp stream below me. Now my breathing sounds as if someone is breathing for me. I paddle. I make arcs through the water with my hands. Tiny sculpins wink in and out of battered fronds. As I swim along a rock outcrop, I look for seals. I glide along rocks and quiet my movements, searching the sandy bottom. My own dark shadow blots out the light above me. I'm hungry. I search the whole island's submerged perimeter.

Sometimes, it seems I'm getting closer to the water.

A friend of mine kayaking in the sound met Mike once and asked him if he knew me. Mike chuckled, said, "I hear her on the radio . . . She's wondering where the whales are." Mike died four winters ago. The last time I saw him, he had to breathe from an oxygen bottle.

According to traditional stories from all along Alaska's coastline, when killer whales come into a bay, someone will die. A Sugpiaq woman from Nanwalek told me why. "When killer whales come near the village, they're calling someone to join them, so we're sad. A week or two later, someone dies."

There's a killer whale we've named Jack, after the late Jack Evanoff, of Chenega Bay. His niece, Mary, told me that when Jack was an old man, he always said he'd come back as a killer whale with a partially bent over dorsal fin. People told her that they'd seen such a whale out there. I've seen him. There's only one. He's a salmon hunter from a pod that spends more of its time in Prince William Sound than any other. His fin curls around; from the back it looks like a question mark. Lars calls him "Captain Hook."

Sometimes, anchored up in a storm in a place called Pony Cove, I joke with Lars about killer whales, make up crazy stories about what they do. I tell him a story:

Long ago, a man from Nanwalek followed some killer whales in his kayak. He thought they might lead him to seals. The whales dove at the head of a bay and disappeared. When the man paddled to shore, he saw human footprints leading into a cave. He followed them. Inside, he saw humans putting on killer whale skins. Once, I tell Lars, humans and animals spoke the same language.

Can science teach me this language?

Science teaches me that there's a truth somewhere, that I can find it, that I can listen and hear something. For years, I recorded the sounds of transients. I scrutinized each call on a sonograph analyzer. I scribbled descriptions of everything I saw. I identified hunting calls, resting calls, social calls, long-distance contact calls, but I never deciphered the language of the whale that eats only mammals, that speaks mostly silence. The language of the killer whale eluded me.

What I did learn was that it's not difficult, in the moment, to surrender to not knowing. To be a watcher. Like a transient, who finds its prey by deciphering sounds, to listen.

Once I found an old picture of Mike as a child on the schoolhouse steps at Old Chenega. I recognized him by his big ears. Sometimes, Mike walked with me along the shore of his island. He'd stop suddenly, motion me to be quiet. "Listen," he'd say. "I hear something."

Montague Island's reflection extends a long way down into the water in the afternoon's heavy light. Whales swim along its snowy flanks, across green slopes, skim the tops of conifer stands, along bare rock, then dive down *into* the mountains. When they rise again, they break apart the island's reflection.

Lars drops a stone into the water. We watch it. The deeper we go, the more knowledge resembles a question mark. Who's asking the questions? We listen. We watch the stone sinking. We watch it spiral out of sight. Science. It's like that.

Jo Going

Otter Woman

I shall numb you,
paralyze you
with the patter of paws,
and whistle
stupefaction.

In the dusk
and half light,
rain of an impending
violet,
cedar smoke, and damp moss.

Down the bent passage
worn by the raking of claws,
down past old roots
and rotting leaves,
the smell of matted fur,
wet and musky.

We shall touch
in silence,
while from the world
distant above,
the song of the wood thrush
calling.

Joanna Wassillie

That Which Sustains Us

When I think back to my youth, my earliest memories inevitably include details of the landscape that comprises the heart of the Yukon–Kuskokwim Delta. It is lush, not in the same way a tropical jungle is lush, but in a more understated way. It is as much a player in the drama of my childhood as any of my family or friends. It is, after all, alive, and therefore communicates in its own way with the world around it.

To fully appreciate the diversity of this region of Alaska, you must be prepared to look past the seeming monotony of endless tundra, the countless lakes and ponds. You must take in stride the weeks on end of rain during the summer months, and the ruthlessly arid winter winds. It is a region whose complexity stretches out to include every possible form of water. That is what struck me the most as a child: the ways that we were tied with water. The very base of our Yup'ik Eskimo culture was formed by, and took into account, the resources available to us directly and indirectly by water.

The most obvious connection to me, and the one I enjoyed the most, was that of fishing. Every summer, my family's world, and those of everyone we knew, became centered around the river and what it had to offer. As thousands of generations of Yup'ik before them had done, the men and boys of our village would set out to gather the fish that would sustain us throughout the year.

I lived with my grandparents in a small, one-room log cabin close to the bank of the Yukon River, which swept past our village in various speeds according to the season. In the summer months, once the tumult of spring breakup was over, the river assumed a lazy, yet purposeful, rhythm. It was as concrete a foundation of my childhood as my grandparents.

My grandfather would set out with his small wooden boat, and I would ride in the bow, peering down into the murky depths of the Yukon River. I watched in admiration as salmon, glinting in the sun, were hoisted up into the boat. I carried them up to my grandmother, who cut them to be dried and smoked. We took turns hauling water from the river, which she used to wash off the fish blood and slime from her working table—an old, giant wooden spool, around which electric cable was once wrapped. It was such a comfort to know that, later on, we would savor those very same filets of salmon.

As a college student, I often experienced pangs of homesickness, mostly for the familiarity and taste of traditional Yup'ik foods. Not until I was deprived of these Native foods did I realize how much a part of our culture they are. To be Native is to know the variations of fish; to be able to know the seasons for various species; to know the best means of preparing them for future use. Food even managed to find its way into the ceremonies of our people, playing a role in the parlaying between villages; the sharing of wealth always included food. It had to. Food was the only real form of wealth our people possessed.

Even today, the sharing tradition between my hometown of Pilot Station and St. Mary's (a mere fifteen air miles away) continues. There is a special time during the annual potlatch where the men and boys of the visiting village are invited to the hosts' community hall (*qasgiq*) and are there presented with a variety of frozen fish or perhaps even seal. It is a way of sharing as well as proving to one another what effective providers their men are. Fish. Which comes from the river. Seal. Which comes from that place where all waters of the earth flow: the ocean (*imarpik*). Even in our ceremonies, we are directly tied to the life-giving necessity of the waters.

Beyond the practical sustenance the river gives us, it also provides us the means through which we might search out other resources. We take boats upriver in search of wood to use in our smokehouses and steamhouses (*maqi*). We use the river tributaries to take us to prime berry-picking spots. In the spring, after the river has broken up and the spring floodwaters subside, the banks of the river are strewn with useful things: small pieces of birch bark we can use for starting fires, interesting pieces of worn wood, colorful rocks carried down from places far away.

Yes, the water around us brings us many things, and yet it is so easy to take for granted. There is a sense of confidence, as with other aspects of childhood, that things will always be the same, that things will remain as they were when we first knew them.

Now I know better. It started slowly, with the occasional piece of plastic trash found on the beach—washed from some camping trip upriver, perhaps?

Then, I began noticing garbage (mostly tin cans and pop cans) strewn carelessly among the places people went berry picking or camping. Or shotgun shells, emptied out and used, peppering favorite hunting spots. And there were telltale signs of gasoline and oil leaks on the riverbanks—colorful swirls seeping into the ground. It seemed that everywhere I went, there was evidence of man having been there, of man's use of modern trappings.

It has been fashionable in the past to romanticize the "environmental friendliness" of Yup'ik culture. Just as with the notion of "the noble savage," the good intent of the belief tends to want to point the finger of blame on the corruptive influence of the predominant (i.e., "White") culture. It seeks to glorify a past that, while undoubtedly less destructive to the ecosystem, was certainly not picture-perfect.

Any archaeologist who seeks to study more of past Native cultures, and especially old settlements, will seek out that treasure trove of knowledge: the garbage heaps of the past. (The more academically correct term is "midden," but it's all the same.) People learn about the past from what is thrown away, from what is left to be buried over by generations of dirt, houses, bones, etc.

It is with this in mind that I often walk across the tundra today. What would a detergent container tell the generations of the future? That we didn't foresee any future shortages? Any imminent depletion of our resources? Or will they just consider it for what it is? Junk. Refuse. Things not wanted or needed any longer. Will they have justifiable evidence on hand—or should I say *on land*—to prove that those who were here before them cared nothing about preserving the balance of the ecology? About preserving and protecting the very environment that had supported them?

There are some flaws in the way that we view our relationship with our environment. To be fair, traditional Yup'ik culture did not have any serious ways to destroy our ecosystem. How could we? Everything we made was a direct product of our environment, either from the animals we hunted or the land (wood, earth, plants). Our garbage was not polluting anything, as it readily decomposed and blended right in with everything else in fairly short notice.

Ah. If only it were that easy today. Now, when we discard what is considered "used up" and therefore "unusable," the items are around for a long time. A very long time. They even have the power to be deadly if not disposed of properly. Somewhere along the way, however, our cultures have failed to make the necessary adaptation to our modern relationship with the environment.

No longer will our garbage heaps blend into the ground in a few generations' time. They are there, polluting our landscapes and waters for all to see. It is necessary that we live up to our responsibility to the earth and protect our environment *from* ourselves, *for* ourselves.

I remember as a child there were special places to drink fresh water along the beach in the summer. Someone would dig a hole into the ground, near the rocky cliffs. Not a very deep hole, but enough to dip a ladle or a cup. The water seeped out of the natural water table in the rocks. We could just bring a cup over, drink our fill, and then return to our activities. I would not touch that water today for anything; too many gasoline spills on the beach have made it a risk. I find that sad. It was a special thing, being able to drink water that way. It is only one more connection with my childhood that has not survived into modern times, a relic of bygone days. But does it have to be this way?

Wouldn't it be wonderful to feel free to drink well water again? Wouldn't it be refreshing to enjoy a boating trip and not see pop cans or shotgun shells floating in the water? It is such a simple thing to do, to clean up after ourselves and be more careful. It is such a necessary thing to do if we are to be able to continue with our traditional way of life: living off the land. In order for the land to remain able to sustain us, it must be cared for. The waters must be cared for and protected from pollution. It is the most necessary of our responsibilities as the First People. After all, it is that which sustains us. It is that which provides us with the food we eat, and the water we drink. It is perhaps even the most important part of our identity, for without it, we have nothing else.

Cathy Rexford

The Ecology of Subsistence

No daylight for two months, an ice chisel slivers
frozen lake water refracting blue cinders.

By light of an oil lamp, a child learns to savor marrow:
cracked caribou bones a heap on the floor.

A sinew, thickly wrapped in soot, threads through
the meat on her chin: a tattoo in three slender lines.

One white ptarmigan plume fastened to the lip of
a birch wood basket; thaw approaches; the plume turns brown.

On the edge of the open lead, a toggle-head harpoon
waits to launch; bowhead sings to krill.

Thickened pack ice cracking, a baleen fishing line
pulls taut a silver dorsal fin of a round white fish.

A slate-blade knife slices along the grain of a caribou
hindquarter; the ice cellar lined in willow branches is empty.

Saltwater suffuses into a flint quarry, offshore
a thick layer of radiation glazes leathered walrus skin.

Alongside shadows of a hummock, a marsh marigold
flattens under three black toes of a sandhill crane.

A translucent sheep horn dipper skims a freshwater stream;
underneath, arctic char lay eggs filled with mercury.

Picked before the fall migration, cloudberries
drench in whale oil, ferment in a sealskin poke.

A tundra swan nests inside a rusted steel drum;
she abandons her newborns hatched a deep crimson.

Sherry Simpson

The Middle Ground

I LAY BELLY DOWN ON A PLATTER OF ROCK and dipped my hand into the stream's surging waters. I was thinking shades of chilling blue. Aquamarine. Turquoise. Azure? But there is no word for distilled sky. The best I could do was to imagine the tincture of Aqua Velva aftershave, the exact odd color of Sheep Creek.

A shout from my friends roused me. We had just begun our hike into the valley, and already I was distracted. Kathy and Kim each grabbed a hand to haul me up the slippery bank. We were not exploring this mountain cleft for pleasure, exactly. Juneau's coastline is embedded by other, more beautiful, valleys. But this place was different. We had come to look carefully, to appraise loss, to affix meanings. We thought it would be the last time we'd see the valley.

It hardly matters now that we were wrong.

On that rare clear day, Sheep Creek Valley unfolded before us like a magician's trick. Snow still thatched alpine slopes, and ridges crimped the sky. Five miles north lay Alaska's third-largest city, Juneau, but you would not guess that while walking on the toes of Mt. Hawthorne and Mt. Roberts.

A dark-haired man wearing a red bandanna around his throat hailed us from beside the stream. He was helping a boy with a fishing pole. Ten years before he had hiked to this spot while hoisting a fifty-pound pack, he said. "Nothing here but mine ruins then," he added, lofting his voice over the stream. "Now there's a mining outfit up there."

As if it were news to us, he said a gold miner from Echo Bay Alaska Inc. had told him that protests wouldn't matter—the dam would go in for sure. The man jerked his thumb at the boy, who was about seven, the same age I was when my family moved to Juneau. "I figure this will be a momentous occasion for him, something to remember when he's old," he said. The boy, intent on rigging his pole, never turned around. "We're going to catch the last trout!" the man shouted as he waved us on.

The season had stalled just after high summer. False hellebore slumped by the trail in brown, gloopy streaks. Chest-high ferns feathered in the breeze, and fireweed extinguished itself in cotton batting. It is hard enough to memorize something as blunt as a mountain, as contained as a stream, much less all this fine, rampant detail. But I looked closely, and I wrote things in my notebook—"drowsy smell of vegetation" and "elderberry clusters" and "devil's club"—as if words have ever been sufficient.

Words were all I had. I covered mining for my hometown newspaper, thrashing my way through jargon-choked thickets of environmental impact statements, consultants' studies, agency reviews. I attended public meetings where nobody ever convinced anyone of anything because everyone was preaching to the choir and calling out for amens. I listened to environmentalists and took notes, and then I called up mine officials and took more notes, and finally I typed out absurd observations for the public, such as: "Figuring out what the mine would mean to water quality, fish, wildlife, and marine organisms won't be easy." What else could I say? I was only the translator, and not a particularly good one.

I told people I had not decided whether I favored or opposed the mine. That was my job, and it seemed true. Juneau's white history is the history of mining, after all—a history that itself has obscured the continuing presence of the Auk Kwaan, the local clan of Tlingit Indians who settled this coast long ago and established fish camps and smokehouses on salmon creeks such as this one. Sheep Creek Valley was named by the town's founding prospectors, who recognized gold but not mountain goats. For fifty years, three of the world's largest gold mines operated in Juneau's backyard, along with many smaller outfits. The dam was part of a plan to reopen the old A-J, a project that promised more jobs and less reliance on oil, government, tourism. Nothing new about this story.

The three of us ambled through groves of old black cottonwood trees, stately and sweet-smelling. Southeast Alaska lies almost completely under the dominion of spruce and hemlock trees, and almost nowhere else around Juneau do cottonwoods thrive as renegades from the rain forest. Environmentalists argued that the trees made Sheep Creek Valley "unique" and hence worth saving. It was a small irony that these trees probably would not exist if not for the first miners who cleared the valley floor in the late 1800s.

I found myself inventorying signs of inhabitation and wilderness. Here, a rusting shack carpeted by porcupine droppings. There, a giant platform bulldozed from waste ore, engulfing cottonwoods that now rose from stone. I counted leaning telegraph poles, noted obscure bits of corroded metal, listened absently to mechanical rumbles echoing across the valley. Close by, a varied thrush struck the single bell hidden in its throat.

Nobody saw the same things in Sheep Creek Valley; nobody knew how to speak about what they saw. So many of the phrases and words enlisted in the arguments demanded quotation marks to demonstrate their multiple, shifting meanings, depending upon who was speaking. One environmentalist expressed typical sentiments in a letter to the editor urging the company to dump the tailings in "another much less valuable valley." She meant a valley farther from town, one we wouldn't have to look at, but her suggestion also meant enlisting a valley that was sure to be wilder than Sheep Creek. Pro-mine factions used their own rhetorical topology: the valley's history meant it was no longer "true" wilderness, so damming it would mean no great loss in a landscape reamed with thousands of "pristine" valleys. A mine consultant said: "There is a whole lot of emotion swirling around this project. It would be nice if it was based solely on science," as if science is the language that would purify the discourse, making everything simpler for everybody. I could see why he preferred numbers, though; words like "mitigation" and "monitoring" and "reclamation" sprang from a deviant vocabulary far removed from the green actuality of this basin.

All through the valley, life flickered at the edges. We scanned the brushy slopes, looking for black bears. Crossing a quiet slough, we startled a Dolly Varden, a shimmering, seagoing trout named for a Charles Dickens character. The perfect, tiny track of a Sitka black-tailed deer inscribed moist earth. Where Sheep Creek spreads itself into shallow riffles, we paused to watch two plump water ouzels flitting along the bank. One of the slate gray birds marched into the stream until completely submerged, and I longed to see it striding underwater against the current, a bird comfortable in two mutually opposing worlds.

Just before the trail scrabbled up the mountain, we paused by the creek. Kim dipped bare feet into an eddy and gasped at the chill. Kathy glassed a cliff for mountain goats. I contemplated a fire ring littered with empty snack cans of Libby's Peaches Light, trying to calculate the gap between discarding an aluminum can and discarding a mountain of tailings.

I had been afraid to care about this place too much, the way you might hesitate making friends with a terminally ill person. You knew it would be meaningful, that perhaps you could do some good and even learn something, but you also knew that, in the end, you would reap pain and loss. It was too late, of course, but, looking at those Libby's cans, I realized I was against the mine and had been all along. The journalist's creed of objectivity made a convenient suit of chain mail fashioned from empty words carefully chosen for neutrality. It had become so easy to hang back, to let others squabble on the opinion page and in public hearings. I had made a career of weighing pros and cons and forever fiddling with the scale, trying to make it all balance evenly. The middle ground—that was my territory.

We walked back faster than we came. The sun had dipped behind Mt. Roberts, and blue shadows avalanched behind us. The stream seemed unfamiliar from this direction, and so I was startled when we again encountered the fisherman, now setting up a red dome tent beside the creek. He looked up as we passed and shouted, "We caught eight trout!"

We climbed the final ridge that separated us from the rest of the world. The throaty percolation of Sheep Creek fell away behind us. As we rose above the valley, I thought, I should look one more time, but I could not. We crested the ridge and descended into the darkening forest without turning back.

You could say this is a story with a happy ending. Not because soon after this hike I left newspapering so I could say what I wanted to say (I did); not because the whole town rose up against the mine to save the valley (it didn't); but because the price of gold dropped and the company pulled out of the mine project. This was a temporary triumph, if triumph is the right word. Other mines spread like lichen in distant valleys. Forests have been scraped clear of trees on other slopes. Today townspeople argue about cruise ships discharging wastewater and smoke, ecotours clogging the skies with flightseers and the backcountry with people. Mining gold or mining scenery—the results aren't all that different.

And there is, of course, what we bring upon ourselves without protest, without hardly noticing. Juneau's forests, valleys, and wetlands disappear a scrap at a time, paved under by subdivisions and malls. Sheep Creek runs clear, but few other streams survive in their original state; they run through concrete culverts, or dribble through trailer courts, or drain sluggishly through silt and garbage. It must have happened this way in other places. One day you thought there was enough land, or wilderness, or whatever you call it. And on another day, you realized there isn't.

Sheep Creek taught me that knowing your own heart is not enough to save the world. I confess I don't know what is. But if language is all you have, then what else can you use?

Sometimes I imagine lying beside that stream of unnameable blue, face pressed against the damp earth, listening to the low song of the stream humming against the granite floor. Days and nights would rain down, burying me deep in the abyss of history, thinning me into a sedimentary layer of compressed possibilities and place. The sun slanting over the mountain's brow, the warm stirring air, the endless prayer wheel of the stream's passage, and life greening all around—this is what exists beyond measure, beyond calculation, beyond words.

Peggy Shumaker

Chatanika

High, the Chatanika,
high this year, surges
the flats, soaks
the valley. Chatanika

spreads wide
where gravel braids.
Where banks
snug close,

where rock,
earth and root
gang up, high water
scours, carves,

its own image
changeable.
Chatanika, in pools
deep green, in eddies

steeped tea, freezes
and thaws, makes its way on,
full of grayling
flashing like thoughts

among the millions
of mirrors at Minto.
What brought me
exactly here?

Is my flowing
through the world
a fit gift? Have I nourished
more roots than I undercut?

Libby Roderick

Ted and Me

In the long term, the economy and the environment are the same thing.
If it's unenvironmental it is uneconomical. That is the rule of nature.
—Molly Beatty

An enemy is a friend whose story you have not heard.
—Quaker saying

I respect Ted Stevens. He's a sincere man. He has worked extremely hard for his vision of Alaska, a vision of jobs for every Alaskan.

I respect him even though he calls me an extreme environmentalist, and shakes his fist at people like me who work to block oil drilling in the Arctic National Wildlife Refuge. He rages at us on national television, from the floor of the United States Senate, of which he was until 2006 president pro tempore, the elder Republican statesman from Alaska, and head of the powerful Appropriations Committee.

I respect him even though he agreed with his Republican cohort, former Alaska Governor Frank Murkowski, who ran ads during his 2002 campaign labeling Alaskans dedicated to preserving wild places "enemies of the state."

I respect him, in part, because I have a personal history with him. In the early 1960s, when I was very young, my sister and I played at the Stevenses' large log house, a few blocks down from our matchbook-sized home in Anchorage. Ted and Ann Stevens had five small children—Susan, Beth, Walter, Ted, and Ben—and a black lab named George. My mother used to plan parties with funny, warm Ann, whose death in a small plane crash in Alaska in 1978 took a huge toll on Ted and the family. We took refuge in their home when a world-class earthquake ripped a snake-like crevasse under our house on Good Friday in 1964. As a five-year-old, I remember the excitement of boiling snow for drinking water, listening to the ham radio next door to see whether a tidal wave might hit Anchorage, helping put out the flames when George's tail caught fire after he got too close to a candle during the days-long blackout. During our periodic visits to the Stevenses', Ted would sometimes come yelling and swinging through the chaos. I was a shy child, easily frightened. I flattened myself against the wall when he raged at us, hoping to shrink myself down below noticeable levels.

I am less shy now, and less easily frightened. And I find myself trying to figure out the mind and heart of this man who takes environmental efforts in our mutual home state as personal attacks on his character. This man who introduces bills to circumvent the public process on logging the Tongass National Forest, to build roads through the Izembek Wilderness, to place oil rigs in the Arctic Refuge. This man who embodies and expresses many of the values and attitudes we must be able to displace, change, or effectively respond to if we hope to protect the integrity of Alaska's extraordinary lands and waters.

Ted and my father practiced law together in the early 1960s, when Anchorage was a town of roughly 80,000 people, not the 277,000 it is today. There were few paved roads then, and some of the only clues that the nation's largest reserve of oil lay hidden under the geologic folds of Alaska's North Slope were being hand-cranked out by my future parents in a windowless basement, scripted into a newsletter that kept a handful of oil companies informed about developments in Alaska's nascent oil scene. A few years earlier, in 1957, Richfield Oil had struck "black gold" in Swanson River on the Kenai Peninsula. Alaska was buzzing with speculators and landmen, whose job it was to score leases for oil giants. My father made a living keeping his ear to the ground (and his hand on a beer in the local bars frequented by landmen), reporting rumors and facts to interested parties.

Despite their considerably parallel histories and, in many ways, parallel paradigms, Ted and my dad rarely agree on anything political. But in the old days, Alaskans didn't consider political differences a barrier to friendship—there simply weren't enough people in the state to draw those kinds of lines. If you wanted friends, you had to stretch your parameters to include just about everyone, as you still do in many Alaskan communities. In November of 2003, my father attended Ted's eightieth birthday party, celebrated in the halls of the U.S. Senate, joining highly conservative lobbyists, senators, aides, and invited guests. My dad, the former chair of the Alaskan Democratic Party, was warmly and proudly introduced by the Republican octogenarian senator as his old law associate.

On occasion, in an effort to be more effective in my activism on behalf of Alaska's human and nonhuman communities, I've asked my father to tell me where he thinks Ted is coming from on environmental issues. "Ted thinks environmentalists want to shut industry down entirely," my dad reports. "All he thinks about is providing jobs, jobs for Alaskans. Ted thinks people like you are, inexplicably, blocking economic security for Alaskans."

I've thought about that response, long and hard. I love this land; its beauty, power, and wildness lie at the core of my being, buried deep within my body. I want to find ways we can live here and maintain its integrity, somehow merging a vital economy with a wild ecology. Every successful argument against ensuring the health and vitality of the land, waters, air, forests, and wildlife, in Alaska and elsewhere, seems to rely upon a false dichotomy—jobs versus the environment. We are told that our children can either have cheap electricity or a world free of nuclear waste, capital growth or calving caribou. Remarkable numbers of Americans stop thinking in the face of this argument and agree to continue to devastate the biosphere on which our lives and economies literally depend in order to remain "secure." So this argument needs a louder response than we "alarmists" (as a Bush spokesperson recently dubbed environmentalists) have yet given it. Because we, too, have children who need to eat; we, too, need and want livelihoods that put food on the table. And, with the population we currently have, not everyone can live directly off the land. But, unlike our friends on the other side of this argument, we see dangers where they see only opportunities, and limits where they see virtually none.

My dad also thinks Ted is railing less at Alaskan environmentalists like myself than at what he perceives as "wealthy people in the East who are preventing development in Alaska." If there is a single theme that unites many of Ted's generation of non-Native Alaskans, it is the fight against Outsiders controlling Alaskan interests—Seattle fishing interests that dominated Alaskan fisheries throughout the early years of the state and continue to dominate much of it today; the federal government dictating how things should be done in rural Alaskan bush villages (communities as different from Washington, D.C. as the earth is from Mars); Outsiders who own real estate in Alaska and who, as a result, have unusual influence because less than one percent of the state is in private hands. Considerable funding for efforts to protect wilderness areas in

Alaska does come from sources outside the state, often from wealthy Eastern activists who have seen what logging, drilling, mining, factories, nuclear plants, and other "businesses as usual" have done to the rest of the country. Most of the funding for major oil, logging, and mining endeavors in Alaskan wilderness also comes from Outside sources—some, like British Petroleum, from as far away as England. This funding gets less mention from Ted, in spite of the fact that the majority of the profits and jobs created by such enterprises tend to go to Outsiders. Not to mention the fact that, next to oil, the huge amount of federal money Ted brings into Alaska as our senior senator constitutes the lion's share of the state's economy.

According to the late Danaan Parry, a world-renowned expert in conflict resolution working in the most challenging regions of the globe, the first principle of conflict resolution is that the presenting problem is never the real problem. In other words, the politically polarized positions in which we so often find ourselves locked are merely the tips of icebergs—icebergs of stories, histories, passions, loves, disappointments; of betrayals, hopes, dreams, losses, and unmet needs. We are so seldom asked to tell our stories, to reveal the intergenerational aspirations and heartbreaks buried beneath our positions, that we cannot see the humanity fueling the passion on all sides of the environmental conflict. Stripped of our humanness, we turn each other into strangers: "enemies of the state."

But I cannot see Ted Stevens, or anyone else I disagree with, as the enemy. The Dalai Lama urges us: "Be compassionate, not just to your friends, but to everyone." Demonizing people, no matter what they do, turns our relationships into war zones, and wars have no winners. We can and must oppose positions, challenge attitudes and institutions, hold people accountable for their misdeeds, organize, resist, protest, and do what we have to do to effect change in this society, but we need to do so without dehumanizing our fellow travelers on this fragile planet. To do otherwise makes us embodiments of the very disconnection and violence that allows environmental devastation to occur in the first place. So I have to dig a little deeper into who Ted is and what drives him.

The emotional imprints from childhood often shape adult perspectives. My own Alaskan upbringing left me with an impression of almost limitless land, wilderness, wildlife, tundra, and sky. I thought the whole world was similarly blessed. Humans were small players in the huge expanse of earth surrounding us, and certainly not at the top of the food chain. The rapid, bewildering, reckless damage done to Alaskan lands and communities during the pipeline boom of the early 1970s—the almost overnight appearance in my hometown of overpasses, Taco Bells, huge houses on once wild hillsides, street drugs, strip malls, and pavement, pavement, pavement—traumatized a teenager who had yet to realize that she was an American, not just an Alaskan, citizen. The Prince William Sound oil spill of 1989 almost killed me. I have toured the North American continent for years, speaking and singing on behalf of environmental and human sustainability. But, naive child of Alaska that I was, it wasn't until 2002, when I sang at a conference on the restoration of fisheries, forests, and the aftermath of fires in Oregon, that I had ever considered the possibility that my beloved Alaskan wilderness might truly be in serious danger. This awareness hit me when another conference leader made a casual comment about the fact that ninety-one percent of the wild salmon runs in the Northwest waterways are gone.

The wild salmon are almost gone. In the Pacific Northwest.

Salmon are, to me, a totemic creature. They are the mother of all foods, the center of much Native sustenance, the canaries in the mine for the wilds and for the world. An Alaska without salmon is an unthinkable place. And, interestingly, Ted's best acts of conservation have dealt with preserving wild fisheries for Alaskans. He, too, does not want a repeat of the depletion of fish suffered by our compatriots in the lower forty-eight states. The disappearance of salmon, like the decimation before them of buffalo, mountain lion, grizzlies, and the languages and cultures of the Mohican, Yahi, Natchez, and other indigenous nations, shadow the westward emigration of North Americans. And after the Pacific Northwest, there is nowhere else to emigrate but "north to Alaska."

Ted was raised by an aunt and uncle, and his childhood undoubtedly left an emotional imprint on him as well, possibly marked by a powerful need for security, a job. Possibly scarred by the scarcity of the Great Depression. Surely marked by the dominant American version of manhood that dictates that a man support his family solely on the strength of his own efforts, an increasingly impossible task.

The lands of Alaska often represent hope for working-class people from the Lower Forty-eight trying to make it in America's supposedly classless society, in which the odds are rigged from the beginning towards a few privileged groups and families. The fierce individualism and suspicion of authority that have shaped this country since its founding have led generations of American workers to move farther and farther west to try and make it alone. Ironically, "making it alone" in the United States has almost always involved heavy reliance on everything from government subsidies to stolen indigenous lands to unpaid slaves to underpaid immigrant labor. We could work hard to form truly effective labor movements like the late-nineteenth-century Populist movement, in which hundreds of thousands of Americans fought against monopolies and worked to place limits on corporate greed and the concentration of economic power. We could demand that our government provide truly safe social and economic safety nets like many of our Western European counterparts. Instead, most U.S. citizens try to provide security for our families all by ourselves, even though it leaves many of us deeply insecure, both materially and psychologically. And, historically, our favorite method has been moving west in search of gold, oil, timber, jobs.

These jobs have, for the most part, been nonrenewable, fossil fuel–based jobs. For men like Ted Stevens, the very meaning of a job is one in these extractive industries. Those of us standing on the other side of the argument with Ted see this as an old paradigm. And an old definition of a job. These are not the jobs that offer a sustainable future for Alaska or for the world. The new jobs require different investments, different infrastructures, different winners and losers in the new economy, which is one of the primary reasons our ideas generate so much resistance. But it's not just vested interests that create the conflict. Many Americans of Ted's generation are, understandably, having a hard time retooling their brains to make room for these new definitions, this new picture of what an economy is going to have to look like if we are going to have an economy at all.

And, until fairly recently, the environmental movement has not provided much help. If Ted's cohort of leaders has largely ignored the environmental part of the economy, many of us in the environmental movement have skirted the economic part of ecological sustainability. We choose to stay out of the hard discussions about what a new economy will actually look like, how we will get there, and the class and other inequities that will have to be addressed to make that transition.

A number of years ago, as an activist against the development and spread of nuclear weapons, I was introduced to a project based in Oregon entitled Imagining Peace. The initiative was launched in response to the fact that, while most people have a clear picture of what war looks like, they have only the vaguest notion of what peace might actually entail. To counter the popular sense that peace is simply a lull between wars, Imagining Peace conducted events around the state asking people to draw pictures of what healthy, peaceful communities would look like. Participants began by depicting the violence-ridden, pesticide-drenched, poverty-plagued societies all around them, followed by illustrations of the communities they would like to see. Based on these images, Imagining Peace put together a jigsaw puzzle that showed the current images on the left side of the puzzle, the pictures of peaceful communities on the right, and transitional images linking the two sides together, so viewers could begin to imagine the steps we need to take to move from a violence-based world towards a more peaceful one.

Likewise, we need to reimagine our economy and the economic transitions necessary to achieve a sustainable global future. Although many Alaska Natives suspect it of being just another term for White folks doing business as usual (keeping our four-thousand-square-foot houses and SUVs, simply running them on more efficient fuels), "sustainability" as a movement and a term is catching fire across the United States and the world. The word is intended to encompass not only the need for environmental integrity, but also the protection of economic and social vitality and viability. It expresses the growing awareness that the future—if our grandchildren and their grandchildren are to have one worth living in—will truly have to be a world that works for everyone. A world that works not only for environmentalists, but also for loggers and miners who have been retrained for sustainable forestry and mining our landfills for metals; not only computer programmers in Silicon Valley, but former sweatshop laborers in Taiwan; not only middle-class White folks, but also working-class African-American and Latino and Chinese and Laotian and Senagalese folks. And not only humans, but also eagles and moose and grasslands and wetlands and old-growth rain forests and salmon. Our jobs and economies aren't going to mean much if we can't breathe the air, drink the water, and find salmon in our rivers to feed our families. In fact, if we continue as we have been, we may not even have an economy to speak of, since all economies are based, at the root, on what we take from the biosphere, and we are rapidly taking much of it and replacing little. But our natural environment won't be preserved either if we don't address how people will survive within it.

So this is where economics and environmentalism, and Ted's and my views, converge. Self-interest. I want jobs for Alaskans too. I'm an Alaskan. And I need to put food on the table.

Ted and I differ on how to achieve this end. I want to reroute the subsidies currently going to ConocoPhillips and British Petroleum and other oil industries into solar and hydro and geothermal and biomass and wind and fisheries. I want to convert places like the old Alaska Seafood International plant not into a casino, as was proposed, but into a plant to recycle glass and aluminum and steel and paper within Alaska so we don't have to ship our waste materials Outside or clog our landfills with usable resources. I want to develop energy-efficient public transportation and buildings and offer incentives for consumers to buy energy-efficient appliances and cars. I want to support the efforts of local farmers and those involved with "biomimicry" enterprises, businesses that borrow from the evolutionary self-interest of the natural world to make products that could help humanity likewise adapt and survive in changing conditions.

Environmentally responsible jobs are different kinds of jobs, but they are jobs nonetheless. Jobs for Alaskans. And if they require initial subsidization to flower into truly sustainable, job-producing industries, just think oil, think automobiles, think natural gas, think corn farmers, think Boeing. The United States government has always subsidized industries it deems worthy of sustaining. And what could be more worthy of sustaining than our life-support system?

This will not be an easy task, in spite of the fact that we already have the technology to reduce the human footprint on earth by ninety percent. And we already have evidence that people and businesses readily follow market incentives that can unleash tremendous creativity and innovation. But the nonsubsistence Alaskan economy and the jobs it provides have always relied overwhelmingly on extractive industries. Ted is not wrong on that front. And the population, far bigger than it once was, is also far more psychologically and economically dependent on those industries. In order to shift to a more adaptive, sustainable economy, all of us will need to roll up our sleeves and contribute our ideas and our energies to make this transition without dehumanizing those with whom we disagree.

Ultimately, of course, the people of Alaska and the United States are going to have to radically alter the way we do business altogether. And radically reduce the amount we consume. Climate change is already hitting Alaska hard, especially in the villages, and we don't have much time. We're going to have to relearn from indigenous cultures how to live in the adaptive, flexible, give-and-take relationships with the natural world that sustained human beings for millennia.

Social change like that does not happen overnight, unless prompted by ecological and economic crisis, and we're headed rapidly for both. In the meantime and in the aftermath, our best hope is to build bridges between one paradigm and another, and one person and another, wherever possible.

In recent years, environmentally committed individuals and organizations throughout the state and the world have begun applying themselves to the task of how to bring our environmental and economic visions together, how to address Ted's concerns as well as those of his supposed enemies. This is the work all of us need to do, whatever side we think we are on, if we are to reach our mutual goal. And we need to do it with respect for all of life, including our "enemies."

Let's do our best. For Ted, for me, for his children and grandchildren and mine and yours, for all the workers now and to come, for indigenous cultures, for the unutterably beautiful and necessary tidal pools, hanging glaciers, porcupine, salmonberries, heron, cranes, aspen, otters, and Sitka black-tailed deer of Alaska, and for our beloved world entire.

This essay was written prior to the financial investigations of legislators in Alaska. Those inquiries have included possible ethics breaches by Senator Ted Stevens. Regardless of the outcome of the probe, the premise that demonizing our enemies never helps still stands.

Tom Sexton

Denali

Just before dawn, a great bell
waiting for sufficient wind.

No summit today, only
mew gulls over the mudflats.

The down of uncountable
swans to the west, October.

Winter, dark at noon, the light
at the bottom of a bowl.

Heavy snow today. This sound
in the willows, Denali.

At a great distance, glaciers,
the pale leaves of early spring.

Light on the summit at dusk:
alabaster, salmon, rose.

Marybeth Holleman

The Cape

For as long as I can remember, I have been searching, but if you had asked me what for, I would have given you the wrong answer. Not because I was lying, but because I didn't know. It was as if I lacked some essential life compass.

When we were dropped off on an island in the middle of the North Pacific, all I knew was that I had been drawn to this far-flung curve of land for over a decade, ever since I first saw its outline on a map. All I expected from this place that few people visit was to spend time immersed in the wild. Yet there was much I didn't expect, couldn't have expected.

Our pilot, Steve, landed on a wide, ragged shelf of rock between the cliffs and the sea that looked like a hardened lava flow; it was pitted, rough, and sand-blasted, with a single orange flag marking the stretch Steve had cleared of the largest stones. Steve was also long-distance caretaker of what would be our home for a week: a lighthouse station, demanned over twenty years ago and demoted to a minor navigation aid, perched on an exposed cape jutting into the ocean.

Before taking off, Steve pointed out our two-mile route. We couldn't see our destination, but we could see the arc of sandy beach leading to the long, straight stretch of boulder beach and beyond, where it must bend and lead to the cape that has been called the most dangerous point of land along the Pacific coast.

"It's the middle stretch that's the worst," Steve warned. "Are you sure you need all that gear?"

Rick and I pared down our belongings once more, stashing some in the Cessna, while my eight-year-old son, James, collected kelp stems. I helped

James secure the long dried stalks to his pack, so that he looked like a terrestrial octopus, light brown legs the color of his hair unfurling from his back. I was fastening the last stalk when Rick looked up and frowned.

"We don't want to start collecting things yet," he said. "It's a long walk and we'll find plenty at the lighthouse."

After the rock ledge and the gentle slope of sandy beach, we encountered a boulder field that made sense of Steve's concern. The boulders were so slick with algae that I tried to wedge each step between two rocks rather than on top of them. James lost his initial excitement, and, one by one, his kelp legs broke or fell off. He struggled with his small backpack until I strapped it onto my chest.

"I need a break," he said, plopping onto one of the few dry boulders. "My legs hurt."

We traced an increasingly narrow shoreline bounded by the sea and a vertical cliff hundreds of feet high. The soft cliff rock fell in avalanches of pebbles or large boulders. One tan rock ledge was punctuated with house-sized boulders; the cliff face was pocketed with craters where each boulder had let go, and bulging with boulders that, at any moment, might do the same. So we hiked as close to the water's edge, and as far from the cliff, as possible—though it lengthened the distance we had to traverse.

Encumbered and slow, I fell into a trance; all that existed was within the sphere of my arms and legs—slick boulders, the shallow tide pool, the tumble of rock shards. The hot sun and the difficult terrain, which made the slightest glance up risky, seemed in collusion with the shoreline to make me lightheaded, almost hallucinatory. A dreaminess seemed to blanket us all. James no longer asked to stop, but moved among the rocks, stopping to crouch by a pool of water, picking up a stick and dragging it along. Rick still led the way, but no longer at his usual deliberate pace.

We picked our way through several types of beach, each so different and vast that it was like entering another atmosphere. The ledge with house-sized boulders gave way to a shelf of razor-sharp rock teeth; these receded into a rubble of slate shards, as if a giant had thrown a pot to the ground, shattering it. It reminded me of hiking into the Grand Canyon, where the descent through layers of rock formations was like walking back in time. This was more than just geologic time we were treading back into; it was as if we were dropping into the bowels of existence itself.

Suddenly it appeared: a spire of rock, a green hill, a slender white tower crowned with a dome of windows. The lighthouse. Rising up behind the lighthouse, reducing it to the size of a toy, towered a thousand-foot cliff. Its face rose like a vertical symbol of yin-yang—white as the cliffs of Dover on one side, and polished black on the other. The sea fanned out from the cliffs to the horizon, an expanse of blue broken only by a pyramid of dark granite attached to shore by a thread of sand.

Still, there was something commanding about the lighthouse. Maybe because it was here at all, at the edge of the world, after nearly one hundred years not yet succumbed to the terrible storms of the North Pacific.

I've always been drawn to capes. Each time I visit my family in North Carolina, I feel the pull to spend a few days at Cape Hatteras. That cape is called "Graveyard of the Atlantic" because its ever-changing shoals and fickle currents have sunk hundreds of ships, from booty-laden pirate vessels to World War II warships. There I can stand on the beach and watch the cross-hatching waves, roam the wide swath of sand, and find the weathered remains of unlucky vessels. This cape I now hiked toward, with its fierce storms and icy waters, has been equally dangerous for mariners. Before radar and GPS, this lighthouse was the most important navigational aid for hundreds of miles around.

Capes attract. They always have. Cape Canaveral, Cape Horn, Cape of Good Hope, Cape Cod, Cape Hatteras—these are well-known geographical magnets. They are places where the boundary between land and sea collides with great force, an ecotone that tugs. Migrating fish funnel around them, attracting sea lions, whales, sharks, seals. Eddies form on their backsides, and in these eddies are upwellings where nutrients are brought to the surface, feeding the plankton, which are the first link in the marine food chain. Because of this fecundity, and because of the way they jut out into the sea, they are places where cultures and commerce have intersected, collided, and coalesced. Mariners, like salmon, hug their hazardous shores. With their accelerating and reverberating currents and their shifting or rocky shoals, they are navigational nightmares. Mariners once dreaded "rounding the horn" of South America. Protruding into the flat expanse of ocean, they are, like Canaveral, good places from which to take off into the universe.

Capes are places where history is made, where rockets shoot into the sky, where great battles are won, where lives are lost, where the efforts of human civilization are scaled down to their proper size. Capes are places of extremes, of the extraordinary, of the inexplicable.

On the USGS map for this island, in the lower right-hand corner where the index usually is, there was just one line, one I'd never before seen: "There are no roads or trails on this quadrangle."

I loved this. I loved that this wild cape, this abandoned lighthouse station, was ours alone for the week. After dropping our packs outside the door to the wooden shed where we would stay, we ran from building to building, up and down every decaying boardwalk and overgrown path, as if to leave our scent at every boundary. We were the only humans on the entire twenty-two-mile-long island.

All my life, I have sought out time alone in the wild—any bit of wild. As a child growing up in the mountains of North Carolina, my favorite times were those when I climbed up onto the garage roof at my parents' home, sitting shielded by branches of pine and mimosa; when I disappeared into the tangle of the vacant lot next door, saplings and honeysuckle vines forming an arched entrance; when I explored among azalea thickets and meadows at Craggy Gardens; when I climbed Devil's Tower and stood in the fog on a rocky ledge, shrouded from all but the rhododendrons and songbirds and blue-gray Appalachians stretching beyond me. These were the times I felt the most attached to the world.

Though James could have spent all afternoon exploring the cluster of buildings in various states of disrepair—not only the lighthouse itself, but the keeper's house, the tram shed, the supply shed, and a wooden helicopter pad for the Coast Guard's annual lighthouse maintenance—Rick urged us to walk out the long spit to the pinnacle and the Steller sea lion rookery. This weather, he said, wouldn't hold.

Everywhere were signs that this calm, sunny day was surreal, even for June; that this cape more often withstood hundred-mile-an-hour winds and thirty-foot seas. The shoreline was littered with a storm-tossed jumble of logs and buoys, fishing line and plastic containers, jetsam and flotsam, bright yellow and blue and orange against a muted shore; the front line of trees above tide line was low-growing, bonsaied by salt and wind; the windows on the cluster of buildings were boarded up, broken, or both; the lighthouse and every metal handle or hinge was rusted a deep blood red.

James found a computer washed ashore, and Rick took his picture sitting before the sandblasted monitor, his hands on the keyboard. We found a hockey glove, left-handed like another James found where our plane had landed. Later, we learned that a shipment of hockey gloves was lost when a storm blew several containers off a barge. I imagined it: thousands of hockey gloves floating on the sea, like the hands of a drowning man reaching for air, in the storms of the Pacific, all of them spreading out and each of them washing ashore on all the thousands of beaches along the undulating coast, mateless.

Looming larger in our view the closer we approached was the pinnacle, perfect in its terrible beauty. This jagged rock spire, tapering to a sharp point and curved like a shark fin, burst straight up from the ocean. Green life grasped it, sneaking into every crack and fold. Murres, tufted puffins, cormorants, gulls, and kittiwakes nested in its hummocks, crevasses, and bouldered base. Even the island itself at this cape end was a rough rock spine where only a thin layer of green clung.

There's something about ragged edges of land jutting up through the sea, the solid rock of which capes are often made, as if they are messengers or harbingers of some deep earth force. As if they rise from the molten hot core to remind us of what we stand upon: a wild living planet. These capes, these borders where land and sea collide, neither giving way—they are concentrations of planetary power.

In contrast to the keenness all around, the hill that the lighthouse and buildings sat on was soft and green, and blanketed in salmonberry bushes. One morning, as we sat on a wooden deck outside the dilapidated keeper's quarters, Rick spied three humpback whales just offshore. Soon we all saw them: sprays of water rocketing upward, flukes like broad, gray hearts rising again and again. Instead of what I was accustomed to—a brief glimpse while the whales raced in the other direction—we watched the whales for hours, for longer than I'd ever imagined. After we descended the hill, wandered the beach, made a fire, took a few walks—they were still there.

All the while above them soared a cloud of shearwaters, hundreds of them gathered like passenger pigeons must have over seas of prairie. They plummeted into the water, catching fish and splashing, then flew up in great waves, clusters of dark birds that billowed out and furled in like the sighing of the wind.

There was a lightness of being here for all of us, a fecundity of the sea, the land, the animals—and us in it. James collected flotsam and created sculptures. He rescued wood frogs from rain-filled bottles and carved out a pond for them. He pointed to the different colors of seaweed in tide pools, noticing minute details. He picked berries, for the first time showing an interest in them, reaching so far into the prickly, thick bushes that he disappeared, just to grasp the juiciest one.

Yet our lightness contrasted sharply with the way humans had historically recorded this place—as a place to fear, in which to go insane, to see demons and dragons. On the flight out, Steve had told us stories of lighthouse keepers who stayed here before the station was demanned. These keepers were left here alone for months at a time; they lived with several-ton boulders crashing down the rock cliffs from behind, and with storm waves crashing and frothing from the sea in front, some so powerful they blew kelp up onto the light. Their station, a cluster of four buildings and the lighthouse cleaved to a treeless hill between these two great and dangerous forces, must have felt like slim comfort, like too fragile of a shell to contain their lives. One made himself Emperor of the station, handing out impossible orders to the others. When the supply crew arrived, they remarked on his wide, wild stare, and removed him from the island. Another keeper, in order to ease his claustrophobia, rowed a dingy out to sea. He never returned. Others returned home with stories of ghosts and sea monsters. This place tested a person's strength in more ways than one. This place upended the reasons to be here.

After seeing the cape for myself, I wasn't surprised at the keeper stories. This cape was overwhelming, and most of us aren't used to sustained contact with the overwhelming. We have to search for meaning, make connections to our own experience, make tangible sense of what is beyond our senses. The keepers, charged with preventing this collusion of natural forces from harming passing ships, must have struggled hard to make their own fantastical meanings. But I was here with only a desire for contact.

My experiences in the wild when I felt most of the world around me often contained brief moments that I can only describe as being fully alive. Crouched beside a stream, looking up at the mist rising from its surface, I'd be suddenly struck by a deep feeling of union to all around me, with such felt sense that I'd be left in tears, not of pain or sorrow but of joy. These moments, when the boundaries between myself and the yellow warbler, the oak branches, the mist, the ridgetop, seemed to disappear—these I began to seek out. I wanted to feel the contact more often; I wanted a sustained fullness.

Raised in the Catholic tradition, steeped in twice-weekly sermons and Bible readings, I considered these moments of grace. Ever since I was old enough to grasp what our priest said each Sunday, I assumed the God he spoke of was in these moments—the only times I felt the presence of something more than what I could see, touch, taste, smell, hear. I found clues in others' experiences. Virginia Woolf, who called them "moments of being," considered them a "token of some real thing behind appearances." Mary Austin, who experienced them wandering alone in the desert, called them "flashes of mutual awareness," recognizing a consciousness on the part of the nonhuman as well. These moments, these flashes, became my holy grail.

Knowing that I had but one week at this magnificent cape, knowing that it had taken fifteen years to get here, I was more than ready.

We walked the beach. We climbed the beach. Boulders, huge, so that we tried to hop from one to another. Leap-frogging. James and I jumping about like sprites.

We picked up the scent of something foul. We were each making guesses about what it might be when finally we saw it: large pieces of white, formless mush, surrounded by a scattering of huge bones.

A whale carcass. The bones were lodged between boulders, strewn by storms along a wide swath of beach. The white blubber was one huge mass, a formless jellylike puddle at least fifteen feet wide all around, a few bones protruding as if reaching for air. Smaller piles of blubber lay like giant beached jellyfish among boulders. The stench was unbearable; we moved upwind so that we could look longer. The bones carved beautiful, white curves among the dark, angular rock—arcing rib bones, vertebrae of all sizes, and the skull, sitting on top of the largest boulder, facing out to sea.

We weren't the only ones who had discovered this rich death. Hiding under a pile of beached logs was an immature bald eagle. She glared at us with those terrible yellow eyes, that hooked beak—but she didn't fly off. We wondered if she was injured. She was eating off the whale carcass, a good place to be if she wasn't able to fly. But what would happen when storms came? When winter came?

The day before, we had watched a pair of eagles, which were nesting in a large spruce right behind the cabin. The swirl of birds on the pinnacle—puffins, murres, kittiwakes, gulls—included chicks, tasty, flightless chicks. The eagles would swoop in and try to take a chick, then fly back to their nest, right over our heads. As soon as they appeared, the nesting birds swept off the rock, flying around it in great waves, and gulls dive-bombed the great eagles, pecking and screeching loudly. Sometimes they succeeded in driving off the eagle. Sometimes, though, the eagle got a chick, a silhouette of a body and two skinny legs hanging from its talons as it flew back to the nest, most likely to feed to its own chick. Two or three keening gulls followed, and then only one, just to the rim of the dark forest.

Here, life sat side-by-side with death, the two much more entwined than the kind of death, or life, of which we usually speak. Born of necessity and chance, both occurred openly, moment to moment, simultaneously expected, equally accepted. Life and death shimmered in the clear sea air like the beauty of a thousand perfect moments.

As I spent time in places where I might experience those all-too-brief flashes more often and for longer, I learned that such power to create these moments lay in the places themselves. All I had to do was be there and be open, like a lightning rod that, for an instant, felt the jolt of full life. This kind of power has been called *numen*. It's something that can't be photographed or taped or collected, but can be as real and tangible as granite.

Numen is of the natural world, abides by its laws, appears most strongly in what we call wilderness, or in places where humans live in balance with the natural world. The dead whale, the wounded eagle, the stolen chicks were all signs that this place was still so unmediated by humans that we could be here in proper scale. There was a cycle of life here, the eagle feeding on a dead whale, the eagle perhaps soon to be the food of beach hoppers and flies. The mediation of this raw experience, the strong-armed dominion over the wild that characterizes much of human civilization, is what diminishes the numen, makes a place less itself.

We walked only once more out to the sea lions, though every day we heard their bellows. Over a hundred sea lions crowded along the outermost edge of huge rock slabs that were rounded like the backs of surfacing whales encrusted with barnacles. The sea lions' sleek, brown bodies, except for the sheen left by their time in the sea, blended with the rock upon which they rested. The air was pungent with their sea-wet skin and fishy breaths; their bellows and grunts and sighs rose and fell like waves.

A young bull on the shelf nearest us kersplashed into the sea at our approach. A few minutes later, he clambered back out at another spot, this time near an older bull. He was submissive—head down, tilting-bobbing side to side. I imagined that the tiny ears were flattened, too. He pulled out, strong front flippers forward, then a hop-schwunch to bring rear flippers up. But the old bull roared a no, and he slipped back in and porpoised off.

The land was a dangerous edge to them, the way the sea was a dangerous edge to me. For me, the danger was part of the draw. When I'm in a place where physical safety isn't ensured, I pay attention. But the danger of land to the sea lions touched on the faint thread of human history that ran through this cape.

Twenty years ago, thousands of sea lions, not hundreds, had covered these rocks. This cape was a microcosm of what had happened to Stellers up and down this northern coastline: their numbers had dropped by ninety percent;

they were listed as endangered; there was no sign of recovery. While natural causes might also be at fault, for decades fishermen bombed rookeries and shot them on sight for catching fish and ruining nets, and a growing fleet of factory trawlers now scoop up large quantities of the same fish sea lions eat.

I wondered how the disappearance of so many sea lions—their bodies and movements and voices—diminished the numen of this cape. They were so entangled, individual lives and numenosity, so one and the same.

It's said that numen can be restored. Pilgrims in Tibet strengthen the numen of their sacred mountain by approaching it on their knees. Here at this cape, our experience of the numen might have been intensified by our daylong toil along a treacherous shore. Weeks after our trip, Rick returned in a Coast Guard helicopter and landed right at the cape, hoping to rescue the bald eagle. It had disappeared. Later that summer, though, Steve saw it again. Perhaps the numen at capes is harder to destroy; perhaps their knife-edged bulkheads make them more protective, inviolable.

This apogee we stood upon was named after a saint—as if such naming might spare the namer from the named. I wandered around praying the entire time, not to be spared, but to be here. We were all three like apostles ready to follow the divine out across the waves.

At low tide, we explored tide-pool flats that stretched like a multicolored carpet on the western side of the spit. The large, shallow pool was a garden of pink and red coraline plants, bright green sea lettuce, olive green and orange popweed, and the long brown-red fronds of kelp waving gracefully against the surf, a sunrise of seaweed. In shorts and sandals, we scrambled over the cobbles and entered the water, toes first. At first James was reticent, telling me it wasn't warm like Hawaii. But then he ventured a toe and exclaimed, surprised, "It's warm!" Soon he was splashing up to his waist.

We meandered from tide pools out to the other side of the pinnacle. We scaled a huge boulder and sat on dark rock warmed by sun. James climbed down to throw pebbles into waves but I sat, startled by a singular rock rising above the waves covered with sea stars in a rainbow of colors. Rick climbed down with James and still I stared, mesmerized by the mosaic. Behind me the cliffs vibrated with birds and sea lions, below me the waters brimmed with fish and whales, and beyond, the sky thrummed with the wingbeats of birds. Rick and James had to pry me from that boulder, like a limpet at low tide.

Later, I left them tending our daily beach fire and hiked once more to that rock. I felt as I did on West Virginia's New River Bridge twenty years earlier, standing next to a sky diver, watching him jump and feeling like I might jump next. Only this time I didn't step back. I wanted to dive into that sea, become sea lion, move with that grace, the cool waters sliding over my perfectly smooth skin. Dive off the boulder, arc out and in like a bird after a fish, like the shearwaters that dance with the waves.

In my hand I carried a thick copper lid I had found among beach rocks. The weathered, green copper sat heavy in my palm. I glanced down at it, turned away from the sea, and lumbered back down the beach, contained, awash in a confusion of joy and grief, walking that edge.

Rudolf Otto, an early twentieth-century German philosopher, defined religion in terms of the numinous. The experience of the numinous, he said, is an awareness of the divine that clarifies our sense of our own creatureliness. He called this power, this sensation, *mysterium tremendum et fascinans*—a mystery awesome and entrancing.

There's a desire, once such divinity is experienced, to remain in that holy light. To sustain contact with the overwhelming. To dive into the waves. But there's also a concurrent and essential awareness of our human-ness. This is what grounds us even as we are filled with awe. This is what brings tears to the eyes, what makes us feel even more fully the contact of flesh and bone to rock, tree, soil, air, water. This is what I experienced on the outer rim of the cape, and what, in one swift jolt, shed light on something that had happened during our first few hours here.

After we had walked out to the sea lions that first afternoon, we climbed partway up the pinnacle, just above high-tide line and onto the grasses and mosses that softened sharp rock. All around were wildflowers, wind, sun, the cool breath of the sea. I sat on a ledge and watched James play nearby. Then I felt a hand on my shoulder, and a rush of emotion like a sudden breeze. I turned, expecting to see Rick, but there was no one there. Still, the shadow of the touch lay heavy on my shoulder, undeniable, as real as the rock upon which I perched.

Steve had told us one other story on the flight out. He said the four keepers who were here during the 1964 earthquake were climbing this rock spire when the quake struck. The cape and the spire suddenly thrust upward fourteen feet; one man fell and broke his leg. The other three tried to save him, but they all ended up in the freezing water. The three made it, but the broken-legged man drowned.

Was it he who touched my shoulder? Was I, like so many keepers before me, another character in some ghost's story? Or was I just tired from the taxing hike, imbalanced by this strange new territory, imagining things?

As the days went by, I did sense a presence in this place, but it was broader than any one life or time. I noticed it especially on the cape's outer edge, stronger than the ocean's current: a deep grounding, a pull to stay—the numen of that place, so potent that, as I sat there that first day, on the edge of land and sea and air, I physically *felt* it. Numen, so much more endurant than anything I'd ever experienced, in all my years of searching, that I did not at first believe my own sense.

When we returned home, I made cobbler from the berries James collected. Through the berries we ate, through the water we drank, through the air we breathed, we now carried the Cape within us. Some of that great steady force, running through my veins and muscles and bones. Giving me a measure of power and clarity, eagle-eyed vision, the grace of sea lions under water, peace of humpbacks feeding offshore, synchronicity of shearwaters diving and rising. Through the thousand perfect moments, a sense of my place in this immense world, bathed in the light of the numinous, on the edge of that wild cape. And so bathed, made more human.

John Haines

Return to Richardson, Spring 1981

Somber now, the grizzly hills,
the lake water slack and gray
between the shore and candle ice.

I walk a path under birch and aspen
still leafless in this early April.
Another winter, old neighbors long
departed, and the pole bridge fallen.
I see underfoot, black in the rusty
soil, the leaves of a lost summer.

I remember: it was my hand on the axe
that cleared the trees from this path;
that turned and fenced the garden,
the same hand that split and piled
the cordwood, far back in a time
of grace between the Asian wars.

And I remember the two of us then,
after a long day's work in the hills,
quiet with a book between us, the lamp
turned down, the title long forgotten.

Those words read late in the evening
the pages turned by this hand.
Your voice as you turned to sleep,
and our life like a boat set loose,
going down in the lighted dusk.

It is one more spring in the north.
Over the snow-patched land
a brown wind drives a late flurry
down from the granite ranges.

In this restless air I know,
on this ground I can never forget,
where will I set my foot
with so much passion again.

Mary TallMountain

The Last Wolf

the last wolf hurried toward me
through the ruined city
and I heard his baying echoes
down the steep smashed warrens
of Montgomery Street and past
the few ruby-crowned high-rises
left standing
their lighted elevators useless

passing the flicking red and green
of traffic signals
baying his way eastward
in the mystery of his wild loping gait
closer the sounds in the deadly night
through clutter and rubble of quiet blocks

I heard his voice ascending the hill
and at last his low whine as he came
floor by empty floor to the room
where I sat
in my narrow bed looking west, waiting
I heard him snuffle at the door and
I watched
he trotted across the floor

he laid his long gray muzzle
on the spare white spread
and his eyes burned yellow
his small dotted eyebrows quivered

Yes, I said.
I know what they have done.

Thematic Contents

The following groupings are provided for readers interested in specific regions, flora or fauna of Alaska, or particular issues affecting the state. Some essays and poems are broad and difficult to classify; thus, they may appear under several headings. Other pieces, by contrast, appear under very few.

Geographic

The Moving Out, John Morgan
Walker Lake, Peggy Shumaker

Southcentral

April, Tom Sexton
August, Susan Alexander Derrera
Drink, Molly Lou Freeman
Fall, Shannon Gramse
Ghost Bear, Kaylene Johnson
Going for Water, Karin Dahl
Losing Out to Baseball, Motherhood, and Apple Pie, Susan Pope
October, Buffy McKay
Out of the Depths, Bill Sherwonit
Sweet Spring Grasses, Tom Sexton
Ted and Me, Libby Roderick
The Cape, Marybeth Holleman
The Day the Water Died, Walter Meganack, Sr.
The Experiment, Nancy Lord
The Mighty Sand Lance, Martin Robards
Tree Bonking, Ann Dixon
Two-Part Invention for Winter, Karen Tschannen
Wondering Where the Whales Are, Eva Saulitis
Your Land, Mike Burwell

Southeast

A Voice for Shared Lands, Hank Lentfer
Beached, Pamela A. Miller
Garbage Bears, Hoonah, Ken Waldman
Otter Woman, Jo Going
Slouching Toward Deer Rock, Daniel Henry
The Forest of Eyes, Richard Nelson
The Middle Ground, Sherry Simpson
The Possibility of Witness, Carolyn Servid
Wash Silver, Mike Burwell

Southwest & Aleutians

Cormorant Killer, Jerah Chadwick
Election Day, Anne Coray
Fishing Grounds, Nancy Lord
Light on the Kuskokwim, Alexandra Ellen Appel
Morels, Jerah Chadwick
Precarious Preserve, Anne Coray

Sweet Drug of the Backward Drag, Anne Coray
That Which Sustains Us, Joanna Wassillie

Habitat

Forests

A Voice for Shared Lands, Hank Lentfer
Going for Water, Karin Dahl
In the Shelter of the Forest, Marjorie Kowalski Cole
Losing Out to Baseball, Motherhood, and Apple Pie, Susan Pope
Otter Woman, Jo Going
Partners on the Wheel, Douglas Yates
Return to Richardson, John Haines
The Experiment, Nancy Lord
The Forest of Eyes, Richard Nelson
Tree Bonking, Ann Dixon

Oceans, Seashores, and Islands

Beached, Pamela Miller
Cormorant Killer, Jerah Chadwick
Drink, Molly Lou Freeman
Fishing Grounds, Nancy Lord
Garbage Bears, Hoonah, Ken Waldman
Out of the Depths, Bill Sherwonit
The Cape, Marybeth Holleman
*The Day the Water Died,*Walter Meganack, Sr.
The Ecology of Subsistence, Cathy Rexford
The Mighty Sand Lance, Martin Robards
The Possibility of Witness, Carolyn Servid
What Happens When Polar Bears Leave, Marybeth Holleman
Wondering Where the Whales Are, Eva Saulitis

Rivers and Lakes

August, Susan Alexander Derrera
Chatanika, Peggy Shumaker
Continuing a Conversation on Place, Poetry, Love, Amy Crawford
Fall, Shannon Gramse
Ghost Bear, Kaylene Johnson
If the Owl Calls Again, John Haines
Light on the Kuskokwim, Alexandra Ellen Appel
Precarious Preserve, Anne Coray
Respect Gaalee'ya, Howard Luke

Return to Richardson, John Haines
Slouching Toward Deer Rock, Daniel Henry
That Which Sustains Us, Joanna Wassillie
Walker Lake, Peggy Shumaker
Wash Silver, Mike Burwell

Tundra and Mountains

Bones, Amy Crawford
Crossing Paths, Nick Jans
Curry Ridge, Sally Carricaburu
Denali, Tom Sexton
Due North, Joan Kane
Facing East, R. Glendon Brunk
Finding Refuge, Karen Jettmar
Morels, Jerah Chadwick
Standing on a Heart, Steve Kahn
Sweet Drug of the Backward Drag, Anne Coray
Sweet Spring Grasses, Tom Sexton
The Ecology of Subsistence, Cathy Rexford
The Man Who Skins Animals, John Haines
The Middle Ground, Sherry Simpson
Withdraw, Joan Kane

Human Communities

April, Tom Sexton
August, Susan Alexander Derrera
Dingmait, Joan Kane
Election Day, Anne Coray
Fall, Shannon Gramse
Ice Auguries, Burns Cooper
Light on the Kuskokwim, Alexandra Ellen Appel
October, Buffy McKay
Precarious Preserve, Anne Coray
Respect Gaalee'ya, Howard Luke
Return to Richardson, John Haines
Ted and Me, Libby Roderick
The Last Wolf, Mary TallMountain
The Moving Out, John Morgan
Two-Part Invention for Winter, Karen Tschannen
What Happens When Polar Bears Leave, Marybeth Holleman
Wolf Wars, Nick Jans
Your Land, Mike Burwell

Encountering the Other: Presences

Birds

August, Susan Alexander Derrera
Beached, Pamela Miller
Cormorant Killer, Jerah Chadwick
Denali, Tom Sexton
Dingmait, Joan Kane
Facing East, R. Glendon Brunk
Fall, Shannon Gramse
If the Owl Calls Again, John Haines
Partners on the Wheel, Douglas Yates
The Cape, Marybeth Holleman
The Ecology of Subsistence, Cathy Rexford
The Forest of Eyes, Richard Nelson
The Middle Ground, Sherry Simpson
Walker Lake, Peggy Shumaker

Fish

Chatanika, Peggy Shumaker
Fishing Grounds, Nancy Lord
Out of the Depths, Bill Sherwonit
That Which Sustains Us, Joanna Wassillie
The Ecology of Subsistence, Cathy Rexford
The Middle Ground, Sherry Simpson
The Mighty Sand Lance, Martin Robards
Wash Silver, Mike Burwell

Land Animals

Bones, Amy Crawford
Crossing Paths, Nick Jans
Curry Ridge, Sally Carricaburu
Election Day, Anne Coray
Facing East, R. Glendon Brunk
Finding Refuge, Karen Jettmar
Garbage Bears, Hoonah, Ken Waldman
Ghost Bear, Kaylene Johnson
Otter Woman, Jo Going
Precarious Preserve, Anne Coray
Respect Gaalee'ya, Howard Luke
Standing on a Heart, Steve Kahn
Sweet Spring Grasses, Tom Sexton

The Last Wolf, Mary TallMountain
The Man Who Skins Animals, John Haines
Walker Lake, Peggy Shumaker
Wolf Wars, Nick Jans

Marine Mammals
Fishing Grounds, Nancy Lord
The Cape, Marybeth Holleman
The Ecology of Subsistence, Cathy Rexford
The Possibility of Witness, Carolyn Servid
What Happens When Polar Bears Leave, Marybeth Holleman
Wondering Where the Whales Are, Eva Saulitis

Plants
April, Tom Sexton
Morels, Jerah Chadwick
The Experiment, Nancy Lord
The Middle Ground, Sherry Simpson
Tree Bonking, Ann Dixon
In the Shelter of the Forest, Marjorie Kowalski Cole
The Forest of Eyes, Richard Nelson

Environmental Issues and Themes

Fish and Fisheries
Fishing Grounds, Nancy Lord
Out of the Depths, Bill Sherwonit
The Mighty Sand Lance, Martin Robards

Forestry and Logging
In the Shelter of the Forest, Marjorie Kowalski Cole
Losing Out to Baseball, Motherhood, and Apple Pie, Susan Pope
Slouching Toward Deer Rock, Daniel Henry
The Forest of Eyes, Richard Nelson
Wash Silver, Mike Burwell

Global Warming
Ice Auguries, Burns Cooper
In the Shelter of the Forest, Marjorie Kowalski Cole
Precarious Preserve, Anne Coray
The Experiment, Nancy Lord
What Happens When Polar Bears Leave, Marybeth Holleman

Subsistence Traditions in the Modern World

October, Buffy McKay
Respect Gaalee'ya, Howard Luke
That Which Sustains Us, Joanna Wassillie
The Day the Water Died, Walter Meganack, Sr.
The Ecology of Subsistence, Cathy Rexford

Encroaching Development

Losing Out to Baseball, Motherhood, and Apple Pie, Susan Pope
Crossing Paths, Nick Jans
The Last Wolf, Mary TallMountain

The War on Wolves: Predator Control

Election Day, Anne Coray
Finding Refuge, Karen Jettmar
Ice Auguries, Burns Cooper
Precarious Preserve, Anne Coray
Wolf Wars, Nick Jans

Wildlife Protection and Conflict

Beached, Pamela A. Miller
Cormorant Killer, Jerah Chadwick
Crossing Paths, Nick Jans
Election Day, Anne Coray
Fishing Grounds, Nancy Lord
Garbage Bears, Hoonah, Ken Waldman
Ghost Bear, Kaylene Johnson
Ice Auguries, Burns Cooper
Precarious Preserve, Anne Coray
Respect Gaalee'ya, Howard Luke
Standing on a Heart, Steve Kahn
Sweet Spring Grasses, Tom Sexton
The Man Who Skins Animals, John Haines
Wolf Wars, Nick Jans
Wondering Where the Whales Are, Eva Saulitis

Contributors

ALEXANDRA ELLEN APPEL has lived in Gustavus, Akiak, and Anchorage. She has been a scholar at the Bread Loaf Writers' Conference, and her work has appeared in a variety of literary magazines and anthologies, including features in *PenHouseInk*.

R. GLENDON BRUNK taught at Prescott College. Author of the memoir *Yearning Wild* (Invisible Cities Press), he was a wilderness guide, log-home builder, wildlife biologist, and dogsled racer in Alaska. He wrote and produced *The Last Great Wilderness* multimedia show, which educated thousands of Americans about the Arctic National Wildlife Refuge.

MIKE BURWELL's poetry collection *Cartography of Water* was published by NorthShore Press in 2007. He is also the author of two chapbooks: *North and West* (Heaven Bone Press) and *A Chanting of Waters* (Embers Press). His poems have appeared in *Alaska Quarterly Review*, *ICE-FLOE*, and *Poems & Plays*. Burwell teaches a poetry workshop at the University of Alaska Anchorage, writes environmental impact statements for the Department of the Interior, and maintains a shipwreck database for Alaska.

SALLY CARRICABURU has lived and taught in Alaska since 1975. Her poems have won the *Anchorage Daily News'* and *American Mothers'* writing contests, and "Raven's Nest" was selected by GCI and the Alaska State Council on the Arts in 2002, in conjunction with National Poetry Month. She has published in *Alaska Quarterly Review*, *Interim*, and *Alaska Women Speak*.

JERAH CHADWICK, a twenty-six-year resident of Unalaska, is the author of several chapbooks and a full-length collection, *Story Hunger* (Salmon Press). He guest-edited *Contemporary Art and Writing of the Aleutian Islands* (Penumbra). He is a former Alaska State Writer Laureate.

Marjorie Kowalski Cole has lived in Alaska since 1966, most of that time in Fairbanks. Her poetry, fiction, and essays have appeared in many journals, including *Alaska Quarterly Review, Antigonish Review, Grain, Passages North,* the *Chattahoochee Review*, and *Commonweal*. She is the 2004 winner of the nationally acclaimed Bellwether Award for her novel *Correcting the Landscape*.

Burns Cooper is a linguist who has worked at the University of Alaska Fairbanks since 1990. His poems have appeared in various journals, and he is the author of *Mysterious Music: Rhythm and Free Verse* (Stanford), a linguistic study of the nature of poetic rhythm.

Amy Crawford, was born and raised in Anchorage and currently teaches English in Angoon. Her work has included wilderness guiding and ski coaching. Most recently, she taught English in Mongolia as a Peace Corps Volunteer. She has published in *ICE-FLOE* and other journals.

Karin Dahl is a high school English teacher in Chugiak. Raised on a homestead in Willow, she is a lifelong Alaskan and a former assistant editor at *Alaska Quarterly Review*. Her grandmother was a Sugpiaq from Whale Island.

Susan Alexander Derrera was a finalist for the Ruth Lilly Poetry Fellowship from *Poetry* magazine. Her work has appeared in *Alaska, Alaska Quarterly Review,* and *Our Alaska* (Epicenter Press). She lives and teaches in Anchorage.

Ann Dixon is the author of poems and picture books for children, as well as coauthor of a nonfiction book, *Alone Across the Arctic: One Woman's Epic Journey by Dog Team* (Alaska Northwest Books). She also writes poetry and essays for adults. She lives in Willow.

Molly Lou Freeman took degrees with honors in poetry from Brown University and the University of Iowa Writers' Workshop. Her poetry has been published in numerous American poetry journals. Member of the Alaska artists in residence talent bank, she teaches poetry workshops and English literature at the International School of Paris. She is also editor of *carnet de route*, an annual anthology celebrating design, photography, and cutting-edge poetics in English and French.

Jo Going is a widely exhibited visual artist currently living in Homer. She has received grants and fellowships from the Gottlieb Foundation, the Pollock-Krasner Foundation, the National Endowment for the Arts, and the Alaska State Council on the Arts.

Shannon Gramse cofounded and coedited *ICE-FLOE: International Poetry of the Far North* with his wife, Sarah Kirk. He teaches basic writing at the University of Alaska Anchorage.

John Haines homesteaded near Richardson for more than twenty years. Among his many honors are two Guggenheim Fellowships, the Lenore Marshall Poetry Prize, and a lifetime achievement award from the Library of Congress. He is a former Alaska Poet Laureate.

Jay Hammond was a two-term governor of Alaska, from 1974 to 1982. He published two memoirs with Epicenter Press: *Tales of Alaska's Bush Rat Governor* and *Chips from the Chopping Block.*

Daniel Lee Henry is a lifelong student of rhetoric and communication on the cultural fringe. He has coached high school and university debaters over four decades, including University of Alaska debaters to the national championship in 2002. Author, radio producer, dramaturge, teacher, outdoor adventurer, and remote settler, Henry lives with his psychotherapist wife and ten-year-old son on the roadless side of a bay eight miles from Haines.

Nick Jans is a contributing editor to *Alaska* and serves on *USA Today's* board of editorial contributors. He has written seven books on Alaska wilderness, including *The Grizzly Maze* (Dutton) and *Tracks of the Unseen* (Fulcrum). A twenty-year resident of Inupiaq villages in northwest arctic Alaska, he now lives in Juneau.

Karen Jettmar has lived in Alaska for more than thirty years, working as a park ranger, teacher, environmental activist, freelance writer and photographer, and wilderness guide. She is author of two books on Alaska.

KAYLENE JOHNSON's essays and articles have been published in the *Louisville Review*, the *Los Angeles Times*, Discovery Travel Adventures' *Alaskan Wilderness*, and other publications. She is author of *Trails Across Time* (Kenai Mountains–Turnagain Arm Corridor Communities Association) and *Portrait of the Alaska Railroad* (Alaska Northwest Books). She lives and writes in Wasilla.

STEVE KAHN is a lifelong Alaskan. His essays have appeared in *Alaska*, *Red Mountain Review*, *Pilgrimage*, *ISLE*, and other publications. He is the recipient of an Individual Artist Project Award from the Rasmuson Foundation.

JOAN KANE is Inupiaq Eskimo with family from Mary's Igloo and King Island. She graduated from Harvard College and Columbia University's writing program. Her poems have recently appeared in *Barrow Street*, *Northwest Review*, and *Parthenon West Review*. She was a semifinalist for the 2006 Walt Whitman Award from the Academy of American Poets.

HANK LENTFER lives with his wife and daughter on the bank of a small stream in Gustavus. He is the director of the Gustavus Land Legacy and coeditor of *Arctic Refuge: A Circle of Testimony* (Milkweed Editions). He is currently finishing a book of essays, *Faith of Cranes*.

NANCY LORD is the author of three books of short fiction and three of creative nonfiction. Her nonfiction books, all related to natural history and the environment, are *Fishcamp: Life on an Alaskan Shore*, *Green Alaska: Dreams From the Far Coast*, and *Beluga Days: Tracking a White Whale's Truths* (all from Counterpoint). She lives in Homer.

HOWARD LUKE is a Tanana Athabascan elder from Nenana who began a Spirit Camp on the Tanana River in the 1960s to teach Native traditional knowledge. His autobiography, edited by Jan Steinbright Jackson, is titled *Howard Luke: My Own Trail* (Alaska Native Knowledge Network).

WALTER MEGANACK, SR., was Sugpiaq and Chief of Port Graham, a remote Alutiiq community at the southern end of the Kenai Peninsula.

BUFFY MCKAY recently won a full scholarship to attend the 2008 Key West Literary Seminar and Workshops. Her work has appeared in the *Anchorage Press*, *Anchorage Daily News*, and *Explorations*. She is currently working on a poetry manuscript, and lives in Anchorage.

BILL MCKIBBEN'S most recent books include *The Bill McKibben Reader* (Times Books) and *Deep Economy: The Wealth of Communities and the Durable Future* (Times Books). For over twenty years he has worked to raise awareness about climate change. His newest project is Step It Up 2007.

PAMELA A. MILLER lives in Fairbanks and works for the Northern Alaska Environmental Center, continuing twenty-five years of work protecting the Arctic National Wildlife Refuge as a wildlife biologist, conservationist, oil impact researcher, and wilderness guide. Her arctic photographs have appeared in *Alaska Geographic*, *Alaska's Brooks Range*, *Environmental Science and Technology*, *Science*, *Wilderness*, and others.

JOHN MORGAN has lived in Fairbanks since 1976. He has published three collections of poetry, *The Bone-Duster* (Quarterly Review of Literature Poetry Series), *The Arctic Herd* (University of Alabama Press), and *Walking Past Midnight* (University of Alabama Press). His work has appeared in the *New Yorker, Poetry*, *American Poetry Review*, the *New Republic*, and the *Paris Review*, among others.

RICHARD NELSON is a cultural anthropologist and nature writer, whose books include *Make Prayers to the Raven* (University of Chicago Press), *Heart and Blood: Living with Deer in America* (Knopf), and *The Island Within* (North Point Press)—recipient of the John Burroughs Medal for nature writing. He is a former Alaska State Writer Laureate.

SUSAN POPE migrated from New York to Alaska with her family when she was five. She has published essays in *Pilgrimage* and other journals and collections. She works as a researcher for the School of Social Work at the University of Alaska Anchorage and lives in Anchorage with her husband, daughter, and grandchildren.

Cathy Rexford is Inupiaq of Kaktovik. She is a graduate of the Institute of American Indian Arts creative writing program and the Evergreen State College Native American Studies program. She has been a Gerald Red-Elk Scholar at the Naropa University Summer Writing Program as well as a Truman Capote Literary Scholar. She currently lives in Alaska and is writing her first novel.

Martin Robards is a PhD student at the University of Alaska. He has worked and written as a marine ecologist in Alaska for the seventeen years since leaving his home country of England. He is currently studying the relationship between people and the world they live in. He has published in *Orion* and various academic journals.

Libby Roderick is an internationally acclaimed singer/songwriter, recording artist, teacher, activist, and lifelong Alaskan. She has performed throughout North America, been featured by CNN, the Associated Press, and other national media, been honored by the Alaska Legislature, won many awards, and had one of her songs played on Mars by NASA. Her song "How Could Anyone" has become a global folk classic.

Eva Saulitis's first book, *Leaving Resurrection*, was published by Red Hen Press in 2008. Her poems and essays have appeared in various magazines and anthologies, and she was a contributor to *Home Ground: Language for an American Landscape* (Trinity University Press), edited by Barry Lopez. She lives, teaches, and writes in Homer.

Carolyn Servid is author of the essay collection *Of Landscape and Longing* (Milkweed Editions) and editor of the award-winning anthology *From the Island's Edge: A Sitka Reader* (Graywolf Press). She is also coeditor of *The Book of the Tongass* (Milkweed Editions) and *Arctic Refuge* (Milkweed Editions), and was a contributor to *Home Ground: Language for an American Landscape* (Trinity University Press), edited by Barry Lopez. She lives in Sitka, where she is co-director of The Island Institute.

Tom Sexton was a founding editor of *Alaska Quarterly Review* and its poetry editor for over a decade. He was Alaska's Poet Laureate from 1994 until 2000. His latest collection of poetry is *A Clock with No Hands* (Adastra, 2007).

BILL SHERWONIT is a nature writer who has called Alaska home since 1982. He has contributed essays to a wide range of newspapers, magazines, journals, and anthologies and is the author of eleven books about Alaska. His most recent book is *Living with Wildness: An Alaskan Odyssey*, published in 2008 by the University of Alaska Press.

PEGGY SHUMAKER'S new memoir is *Just Breathe Normally* (University of Nebraska Press). Her most recent book of poems is *Blaze* (Red Hen Press), a collaboration with the painter Kesler Woodward.

SHERRY SIMPSON grew up in Juneau and has lived in Fairbanks, Petersburg, and Anchorage. She is the author of *The Way Winter Comes* (Sasquatch Books) and *The Accidental Explorer* (Sasquatch Books), and her essays have appeared in numerous publications. She teaches creative nonfiction at the University of Alaska Anchorage and with the Rainier Writing Workshop at Pacific Lutheran University.

MARY TALLMOUNTAIN, of Koyukon Athabascan decent, was born in Nulato but raised in California by adoptive parents. Her first book, *There Is No Word for Goodbye* (Blue Cloud Quarterly Press), won the Pushcart Prize. Others include *Light on the Tent Wall* (American Indian Studies Center), *A Quick Brush of Wings* (Freedom Voices), and *Listen to the Night* (Freedom Voices). Active in the Native American literature renaissance, she also taught poetry to children in Alaska villages.

KAREN TSCHANNEN lives and writes in Anchorage. Some of her work has appeared in NorthWest Poets and Artists Calendars, *Alaska Quarterly Review*, *The Sky's Own Light* (Minotaur), and *Love's Shadow: Stories by Women* (Crossing Press).

KEN WALDMAN is the author of six poetry collections, including *Nome Poems* (West End Press), *To Live on This Earth* (West End Press), and *The Secret Visitor's Guide* (Wings Press). He has lived in interior Alaska, southeast Alaska, rural Alaska, and Anchorage.

JOANNA "ANUNG'AQ" WASSILLIE was born in Nome to Roberta Rodin, her Inupiaq mother, and raised in Pilot Station by Yup'ik parents George and Mary Wassillie. She is an educator who has taught in rural Alaska, from

Barrow to Manokotak. She lives in White Mountain with her husband Jack, daughter Katya, and new baby Liam. Her poetry has appeared in *Studies in American Indian Literature*.

DOUGLAS YATES observes the braids of the Tanana River from Ester. A photographer and writer, he works to promote the values found in Alaska's wild places. He has a special interest in boreal forest ecology and the Arctic National Wildlife Refuge. Previous work has been published in the *Christian Science Monitor*, *Alaska*, *Science News*, *Utne Reader*, and *Whole Earth Review*.

Acknowledgments

Unless otherwise noted, all reprints are by permission of the author.

Mike Burwell, "Wash Silver" and "Your Land" are from *Cartography of Water* (NorthShore Press, 2007).

R. Glendon Brunk, "Facing East" is from *Yearning Wild* (Invisible Cities Press, 2002).

Jerah Chadwick, "Morels" is from *Story Hunger* (Salmon Publishing, 1999). "Cormorant Killer" first appeared in *ICE-FLOE.*

Burns Cooper, "Ice Auguries" first appeared in *ICE-FLOE.*

Anne Coray, "Election Day" is from *Bone Strings* (Scarlet Tanager Books, 2005). "Sweet Drug of the Backward Drag" first appeared in the *Southern Review.*

Ann Dixon, "Tree Bonking" first appeared in *SIMUL: Lutheran Voices in Poetry,* in a different form.

Molly Lou Freeman, "Drink" first appeared in *Alaska Quarterly Review.*

Jo Going, "Otter Woman," is from *Wild Cranes* (National Museum of Women in the Arts, Washington, D.C., 1996).

John Haines, "If the Owl Calls Again," and "The Man Who Skins Animals" are from *The Owl in the Mask of the Dreamer* (Graywolf Press, 1993). "Return to Richardson" is from *Of Your Passage, O Summer: Uncollected Poems from the 1960s* (Limberlost Press, 2004).

Marybeth Holleman, "The Cape" first appeared, in a different form, in *Pilgrimage.* "What Happens When Polar Bears Leave" first appeared in *Interdisciplinary Studies in Literature and Environment.*

Nick Jans, "Wolf Wars" and "Crossing Paths" first appeared in *Alaska*.

Steve Kahn, "Standing on a Heart" first appeared in *Alaska*.

Joan Kane, "Due North" and "Withdraw" first appeared in *ICE-FLOE*.

Nancy Lord, "Fishing Grounds" is from *Green Alaska: Dreams From the Far Coast* (Counterpoint Press, 2000). Copyright 2000 by Nancy Lord. Reprinted by permission of Counterpoint Press.

Howard Luke, "Respect Gaalee'ya" first appeared, in a different form, in *Under Northern Lights* (University of Washington Press, 2000). Reprinted by permission of the University of Washington Press and the University of Alaska Museum.

Walter Meganack, Sr., "The Day the Water Died" first appeared as public testimony to the Oiled Mayors of France and Alaska Conference in Kodiak, Alaska, on June 26, 1989. It was subsequently published under different titles in many venues, including *Cries from the Heart* (Wizard Works, 1989) and *Anchorage Daily News*.

John Morgan, "The Moving Out" is from *The Arctic Herd* (University of Alabama Press, 1984).

Richard Nelson, "The Forest of Eyes" is from *The Island Within*. Copyright © 1989 by Richard Nelson. Published by Vintage Books, a division of Random House, Inc. New York, and originally by North Point Press. Reprinted by permission of Susan Bergholz Literary Services, New York, NY and Lamy, NM. All rights reserved.

Cathy Rexford, "The Ecology of Subsistence" previously appeared in *Effigies*, Earthworks Series (Salt Publications, 2007); *Ahani: Indigenous American* Poetry, vol. 9 of *To Topos: Poetry International* (Oregon State University Press, 2007); and *Scrimshaw: Neo-Modern Literature from the Institute of American Indian Arts* (2005–2006).

Eva Saulitis, "Wondering Where the Whales Are" first appeared, in slightly different form, in *Leaving Resurrection* (Red Hen/Boreal, 2008).

Carolyn Servid, "The Possibility of Witness" first appeared in *NIMROD, International Journal of Poetry and Prose*, Spring/Summer, 1995, Vol. 38, No. 2.

Bill Sherwonit, "Out of the Depths" is from *Living with Wildness: An Alaskan Odyssey* (University of Alaska Press, 2008) where it appeared as "A Gift of Halibut."

Peggy Shumaker, "Chatanika" and "Walker Lake" are from *Blaze* (Red Hen Press, 2005).

Tom Sexton, "April" and "Denali" are from *Autumn in the Alaska Range* (Salmon Publishing, 2000). "Sweet Spring Grasses" is from *The Bend Toward Asia* (Salmon Run Press, 1993).

Sherry Simpson, "The Middle Ground" first appeared in *Holding Common Ground: The Individual and Public Lands in the American West* (Eastern Washington University Press, 2005).

Mary TallMountain, "The Last Wolf" is from *Light on the Tent Wall: A Bridging* (American Indian Studies Center, 1990). Reprinted from *Light on the Tent Wall: A Bridging* by permission of the American Indian Studies Center, UCLA. Copyright 1990 Regents of the University of California.

Karen Tschannen, "Two-Part Invention for Winter" first appeared, in a different form, in the February 1993 NorthWest Poets & Writers Calendar.

Ken Waldman, "Garbage Bears" is from *To Live on This Earth* (West End Press, 2002).

We are indebted to Mike Burwell, who provided invaluable assistance at a crucial time in the development of this book. We also deeply appreciate the multifaceted help and support of Steve Kahn, Rick Steiner, and Ann Kahn. Thanks to all the contributors for their patience and powerful words, and most especially and always to Alaska.

About the Editors

Anne Coray is the author of *Bone Strings* (Scarlet Tanager Books); several chapbooks, including *Soon the Wind* (Finishing Line Press); and coauthor of *Lake Clark National Park and Preserve* (Alaska Geographic Association). Her poetry has appeared in the *Southern Review*, *Poetry*, *North American Review*, *Connecticut Review*, the *Women's Review of Books*, and several anthologies. She has been a finalist with White Pine Press, Carnegie Mellon, Water Press & Media, and Bright Hill Press. Coray lives at her birthplace on remote Qizhjeh Vena (Lake Clark) in southwest Alaska.

Marybeth Holleman's most recent book is *The Heart of the Sound: An Alaskan Paradise Found and Nearly Lost*. Her essays, poetry, and articles have appeared in dozens of journals and anthologies, including *North American Review, Alaska Quarterly Review, Orion, Christian Science Monitor, ICE-FLOE, Sierra, Solo*, and *American Nature Writing*. She teaches creative writing at the University of Alaska Anchorage. Raised in the Appalachian mountains of North Carolina, she transplanted to Alaska's Chugach Mountains over twenty years ago.